THE CHEMICAL SOCIETY
MONOGRAPHS FOR TEACHERS No. 32

An Introduction to Enzyme Chemistry

P. F. LEADLAY, MA, DPhil
Department of Biochemistry, Oxford University

LONDON: THE CHEMICAL SOCIETY

Monographs for Teachers

This is another publication in the series of Monographs for Teachers which was launched in 1959 by the Royal Institute of Chemistry. The initial aim of the series was to present concise and authoritative accounts of selected well-defined topics in chemistry for those who teach the subject at GCE Advanced level and above. This scope has now been widened to cover accounts of newer areas of chemistry or of interdisciplinary fields that make use of chemistry. Though intended primarily for teachers of chemistry, the monographs will doubtless be of value also to a wider readership, including students in further and higher education.

First published November 1978

Published by The Chemical Society, Burlington House, London W1V 0BN, and distributed by The Chemical Society, Distribution Centre, Blackhorse Road, Letchworth, Herts SG6 1HN

Printed by Adlard & Son Ltd, Bartholomew Press, Dorking

CONTENTS

Preface

This Monograph is intended as an introduction to the remarkable chemistry of enzymes, which are involved in speeding up almost every essential reaction in living systems. An elementary background in chemistry is assumed, but special terms are defined as they are introduced. The central problem in enzyme chemistry is how to relate the structure of an enzyme to its ability to catalyse a specific chemical reaction. Here, the aim is to emphasise that a reasonable approach to this problem requires information from various complementary experimental methods. Two enzymes are described in detail, to give an idea of the progress which has been made in favourable instances.

The discussion throughout relates to purified enzymes studied in isolation, and in the case of intracellular enzymes their behaviour in the intact cell may well be different. Biologically important interactions between enzymes and other cell components are of great current interest, but, like an extended discussion of the varied chemistry of enzyme cofactors, this falls outside the present scope. Instead, attention is drawn to the growing importance of enzyme technology, that is, the production and use of enzymes for industrial, synthetic and analytical purposes.

I would like to express my thanks to all those who gave permission for reproduction of drawings or figures, and in particular Professors D. Blow, J. Rupley and C.-I. Brändén, who kindly sent photographs or unpublished material. I would also like to thank my wife for typing the manuscript, and several colleagues who read all or part of the manuscript and offered useful criticism.

1. The Enzyme Molecule

Introduction

In 1926 the American biochemist James Sumner obtained a crystalline substance from jack beans which appeared to consist wholly of protein, and which catalysed the rapid breakdown of urea. This was the enzyme urease, and Sumner's work was direct evidence for the idea that urease, and indeed other enzymes, are proteins. At the time his claims met with considerable scepticism, but it is now firmly established that all enzymes are proteins, although in many cases additional components are essential for catalytic activity. These are usually metal ions or small organic molecules.

Proteins are large (macro-) molecules, with a relative molecular mass typically in the range 10 000 to 1 million. They consist of a large number of *amino acid* units or *residues*, covalently linked together to form long, unbranched chains. Formally, proteins are linear polymers (*polypeptides*) with the amino acids as the monomer units, but they differ sharply from their counterparts in synthetic polymer chemistry, in having the different monomer units in a defined order. This chapter aims to show how the number of each type of amino acid in the polypeptide chain, and the sequence in which they are connected together, give to each enzyme protein a well-defined three-dimensional structure and a unique chemical identity.

The amino acids

There are 20 amino acids commonly encountered in enzyme proteins. With one exception, they are carboxylic acids where the adjacent (α) carbon atom is also attached to a primary amino group, a hydrogen atom, and a *side-chain* that characterises the particular amino acid. Their general structure may be written:

$$\begin{array}{c} R \\ | \\ H_2N-CH-COOH \end{array}$$

where R represents the side-chain. The exception is proline, which is strictly speaking an α-imino acid in which the side-chain is looped onto the α-nitrogen atom to form a ring:

proline

1

Acid–base properties

The presence of both an amino and a carboxyl group confers interesting acid–base properties on α-amino acids. An acid is defined, following Brønsted and Lowry, as a proton donor, while a base is a proton acceptor. For the dissociation of a Brønsted acid HA in water

$$HA + H_2O \rightleftharpoons A^- + H_3O^+$$

it is useful to define the quantity pK_a

$$pK_a = -\log_{10} K_a = -\log_{10} \frac{[H_3O^+][A^-]}{[HA]} \qquad 1$$

where K_a is the equilibrium constant for the dissociation. Comparison of 1 with the usual definition of pH

$$pH = -\log_{10}[H_3O^+] \qquad 2$$

gives, after rearrangement,

$$pK_a = pH + \log_{10} \frac{[HA]}{[A^-]} \qquad 3$$

This equation allows the degree of ionisation of the acid to be calculated at any pH, if the pK_a is known.

When the pH is equal to pK_a, the acid is exactly half-dissociated. Thus the pK_a value of an acid expresses its strength directly on the pH scale of acidity, a lower pK_a characterising a stronger acid. The strength of an organic base B is expressed by the pK_a for the dissociation of its conjugate acid, BH^+:

$$BH^+ + H_2O \rightleftharpoons B + H_3O^+$$

Clearly a stronger base is characterised by a higher pK_a for BH^+.

An α-amino acid with no ionisable groups in the side-chain shows two acid dissociations in the pH range 1–13, one near pH 3 and a second near pH 10. There is considerable evidence that such α-amino acids exist, in water at neutral pH, almost entirely in the dipolar form **I** rather than the undissociated form **II**:

$$\underset{\textbf{I}}{H_3N^+\!-\!\overset{\displaystyle R}{\underset{\displaystyle |}{C}H}\!-\!COO^-} \qquad\qquad \underset{\textbf{II}}{H_2N\!-\!\overset{\displaystyle R}{\underset{\displaystyle |}{C}H}\!-\!COOH}$$

The dipolar or zwitterionic structure **I** explains the relatively high solubility of α-amino acids in water, their relatively low solubility in non-polar solvents, and their large dipole moments. It is confirmed by detailed study of their infrared and Raman spectra in solution.

The two acid dissociation processes for, say, glycine are, therefore, best represented:

$$H_3N^+\text{—}CH_2\text{—}COOH + H_2O \rightleftharpoons H_3N^+\text{—}CH_2\text{—}COO^- + H_3O^+ \quad pK_a \ 2.3$$

and

$$H_3N^+\text{—}CH_2\text{—}COO^- + H_2O \rightleftharpoons H_2N\text{—}CH_2\text{—}COO^- + H_3O^+ \quad pK_a \ 9.8$$

In proteins, all except the terminal α-amino and α-carboxyl functions are masked, and the ionisable groups on the side-chains of certain amino acids largely determine the acid–base properties. The chief importance of the acid–base behaviour of individual amino acids is that it provides the basis for several important separation methods. These include electrophoresis, where amino acids migrate under the influence of an applied electric field at a rate proportional to their net charge; and various combinations of electrophoresis with thin layer chromatography, in which amino acids partition between a polar stationary phase and a non-polar moving phase. One method, ion-exchange chromatography, merits a detailed description because of the key role it has played in structural studies on enzymes, but it will be helpful to examine first the properties of individual amino acid side-chains.

Side-chain properties

The structures, names (all of which are unashamedly non-systematic) and usual abbreviations for the common amino acids are given in *Fig 1*. The form shown is in each case the predominant one at neutral pH in water. Inspection of this list shows that the variation in structure of the side-chain is extremely wide. Two proteins differing from one another in amino acid composition, or in the sequence in which the amino acids are linked, may be anticipated to show quite different chemical properties.

The amino acids may be usefully grouped, on the basis of the polarity of their side-chain, into three classes: non-polar, neutral polar, and charged polar. Those regarded as non-polar include glycine and the four amino acids with aliphatic hydrocarbon side-chains, alanine, valine, leucine and isoleucine. Proline, phenylalanine and methionine are also in this class.

The larger the non-polar side-chain, the more hydrophobic (water-hating) character the amino acid acquires. This may be verified experimentally by measuring, for each amino acid, the free energy change on transfer from water into a less polar solvent such as ethanol. As expected, the free energy change is least favourable for glycine, and most favourable for phenylalanine, tyrosine and trypto-phan.

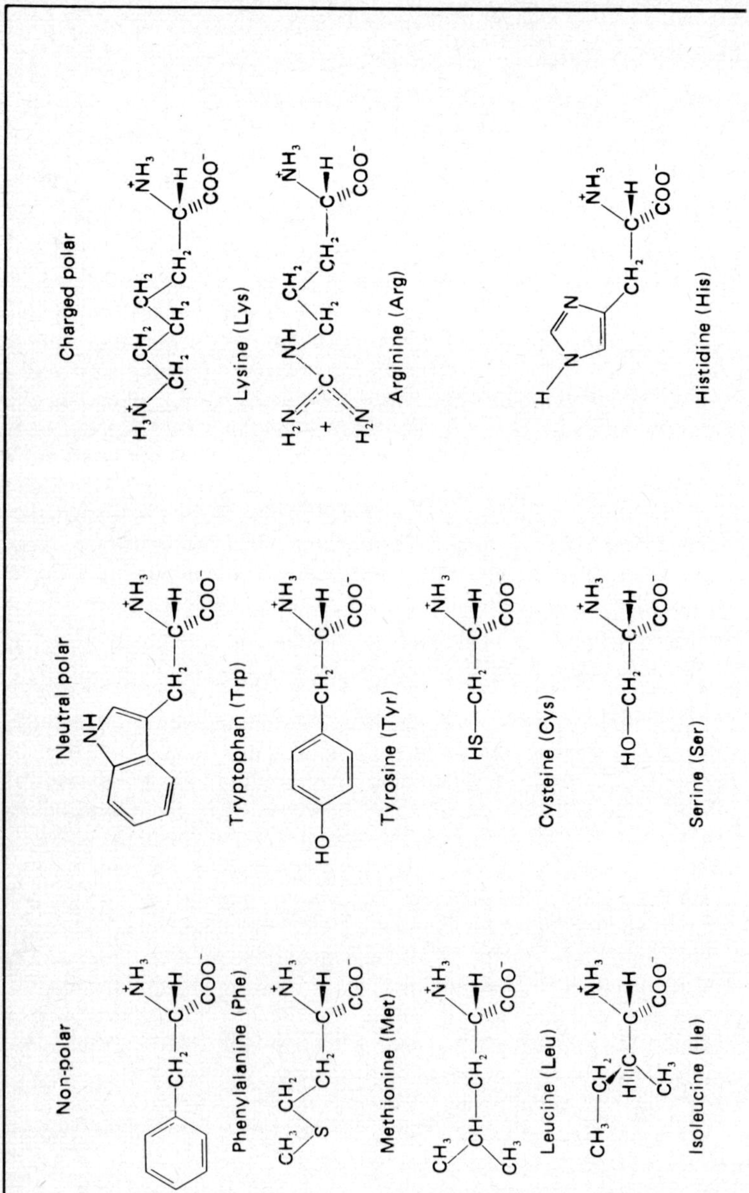

Charged polar

Lysine (Lys)

Arginine (Arg)

Histidine (His)

Neutral polar

Tryptophan (Trp)

Tyrosine (Tyr)

Cysteine (Cys)

Serine (Ser)

Non-polar

Phenylalanine (Phe)

Methionine (Met)

Leucine (Leu)

Isoleucine (Ile)

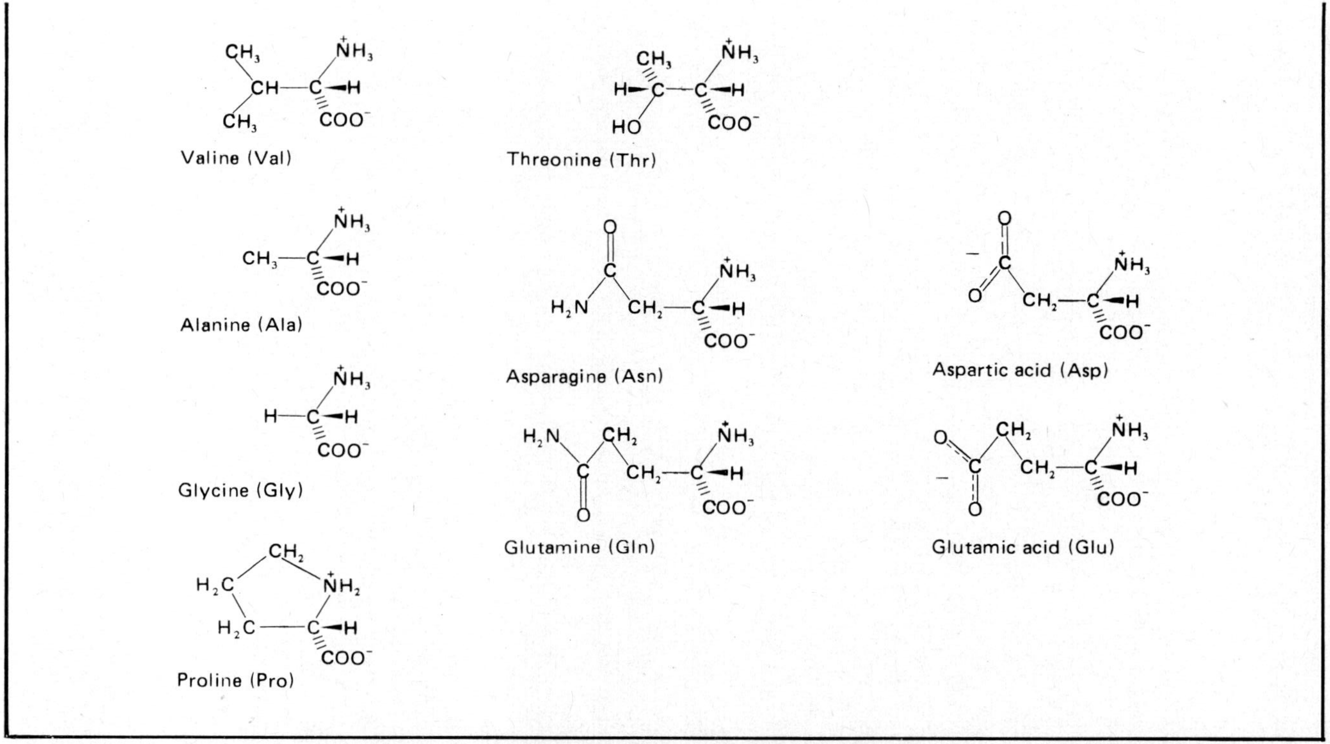

FIG. 1. The 20 amino acids commonly found in enzyme proteins. Dashed bonds are directed below and thickened bonds above the plane of the paper.

The neutral polar class includes the thiol amino acid cysteine, the hydroxylic amino acids serine and threonine, the aromatic amino acids tyrosine and tryptophan (which also have a great deal of non-polar character) and glutamine and asparagine, which have amide side-chains. All these side-chains can engage in strong hydrogen bonding through oxygen, nitrogen or sulphur, to other polar residues or to water molecules in the solvent. Cysteine plays a very important role in maintaining protein structure because of its ability to dimerise, through mild oxidation, to give cystine:

$$\text{M-Cys—SH} + \text{HS—Cys-M} \xrightarrow{\text{[O]}} \text{M-Cys—S—S—Cys-M}$$

Such a cystine or disulphide –S–S– bridge may form between two cysteine residues in the same polypeptide chain, creating a loop, or may cross-link cysteine residues in two separate chains. Note that the –S–S– bridge is much more hydrophobic than the parent cysteine residues.

The charged polar amino acids are those which contain an ionising group, either acidic or basic, in their side-chain. The acidic amino acids, aspartic acid and glutamic acid, carry an extra negative charge at neutral pH, because the pK_a values of their side-chain carboxyl groups are 3.9 and 4.3 respectively. That the pK_a values are different illustrates how sensitively this function reflects small structural differences between acids. The thiol group in cysteine (pK_a 8.3) will be largely in the uncharged form at pH 7, and the phenolic side-chain of tyrosine (pK_a 10.1) almost completely so. The strongly basic amino group at C–6 (the ϵ-carbon) in lysine (pK_a 10.5) and the guanidino group in arginine (pK_a 12.5) will both be completely protonated at neutral pH, giving these amino acids an extra positive charge, while the imidazole ring in histidine (pK_a 6.0) will remain largely (90 per cent) in the uncharged form.

These polar charged side-chains may be expected to exert great influence on protein structure, because of their strong interaction with water and their sensitivity to pH changes. At the same time, the protein environment may profoundly affect the ionisation of individual groups and hence their chemical reactivity. A polar side-chain, at pH values above its pK_a, will be capable of acting either as a Brønsted base, by donating a pair of electrons to a proton, or as a nucleophile, by donating the electron pair to (for example) a suitably electrophilic carbon atom. At pH values lower than its pK_a, the conjugate acid form predominates, and this may function as a Brønsted acid. Catalysis by Brønsted acids and bases is discussed in detail in Chapter 2.

Amino acid analysis

All 20 common amino acids may be cleanly separated from one

another by column ion-exchange chromatography. An ion-exchange resin is used whose matrix contains sulphonic acid groups, which interact strongly with positively charged ions. A column of this material is prepared, and the mixture of amino acids is applied, at a pH (less than 3) where all the amino acids are positively charged and bind strongly to the resin. The ion-exchanger is then washed with a buffer of slightly higher pH, so that the acidic amino acids glutamic, aspartic and cysteic acids (the latter an oxidation product of cysteine), which are least firmly held, are gradually eluted from the column. By steadily increasing the pH and concentration of the buffer, all the other amino acids are eluted in turn. Among the neutral amino acids, the less polar tend to be eluted later, because of favourable interactions with the hydrophobic matrix of the ion-exchange resin. The basic amino acids lysine and arginine are eluted last of all.

This procedure has been refined to the point where it can be incorporated into an automatic system for the quantitative analysis of amino acid mixtures. The principle is shown in *Fig. 2.* As the

FIG. 2. An automatic system for the separation and quantitative determination of amino acids.

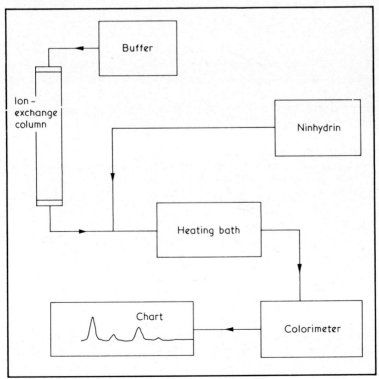

separated amino acids emerge from the column, they are mixed with ninhydrin reagent, which on heating with any amino acid gives— quantitatively—a strongly coloured derivative. The optical density of each peak of colour is measured, and traced out on a chart. Each amino acid emerges in a characteristic position on the chart, and the relative peak areas give a quantitative measure of the original sample composition.

Optical properties of the natural amino acids

The optical activity of any compound, that is the ability to interact with plane-polarised light and rotate its plane of polarisation, is a function of its *symmetry* properties. Two alternative representations of the amino acid serine, conventionally labelled L and D respectively, are shown in *Fig. 3*. Dashed bonds are directed below, and thickened bonds above the plane of the paper.

These two compounds differ only in the spatial disposition of the groups and atoms about the central carbon, and are termed *stereo-isomers*. Although they are related to one another just as an object to its mirror image, they are not identical. Using three-dimensional models, it can readily be shown that no manipulation makes one structure superimposable on the other. Such stereoisomers, which are non-superimposable mirror images of one another, are termed *enantiomers*. With the exception of glycine, all the α-amino acids can exist in either of two enantiomeric forms.

Compounds with this stereochemical property are said to be *chiral*, and to possess chirality (= handedness, from the Greek χειρ, hand). The higher symmetry of glycine renders it *achiral*: it cannot exist in enantiomeric forms. Chirality in α-amino acids is associated with the presence in the molecule of a carbon atom to which four different atoms or groups are attached. Such 'asymmetric' carbon atoms are often described as chiral centres, although strictly speaking chirality is a property of the whole molecule.

Chiral compounds are always optically active, but since one

FIG. 3.   The stereoisomeric forms of serine.

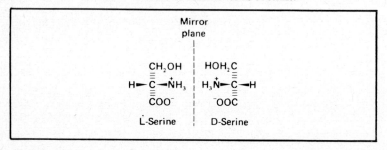

enantiomer will rotate the plane of plane-polarised light in a clockwise sense, and the other by the same amount in the opposite sense, the optical activity will only be measurable if one form is present in excess. Chemical synthesis of an amino acid such as serine will generally provide material which is optically inactive, because it contains equal amounts of both enantiomers. The amino acids obtained by hydrolysis of natural proteins are, however, optically active, because only the L-enantiomer of each chiral amino acid is used in protein biosynthesis.

The peptide bond

The covalent links between adjacent amino acid residues in a polypeptide chain are formed by the condensation of the α-carboxyl group of one residue with the α-amino group of the next:

$$\text{\textbackslash{}V-NH—CH—C—NH—CH—C—NH—CH—C-\textbackslash{}V}$$

peptide bonds

The newly-formed amide bonds are called *peptide* bonds, and the condensation product is termed a *peptide* if the number of amino acid units is small, but a polypeptide or protein if the relative molecular mass exceeds about 10 000. A polypeptide generally contains a free α-amino group at one end, which is usually called the *N-terminus*, and one free α-carboxyl group at the other end (the *C-terminus*). Peptides and polypeptide chains are conventionally drawn with the *N*-terminus on the left. Additional peptide bonds could conceivably be formed involving side-chain amino and carboxyl groups, which would lead to branched or cross-linked chains, but such linkages are extremely rare.

The flexibility of the polypeptide chain is strongly influenced by geometrical restrictions on the peptide links. All the atoms of a peptide bond, including the α-carbon atoms of adjacent amino acid residues, are constrained to lie in a single plane (*Fig. 4*). Ample evidence for the planarity of the peptide bond is provided by extensive x-ray spectroscopic studies on the crystal structure of simple amides and peptides, from which it also emerges that the amide C–N bond is shorter and stronger than a normal C–N single bond. This extra stability is associated with the planar geometry, and there is a considerable energy barrier to any twisting of the C_α–C–N–C_α grouping out of the plane.

A peptide bond could, in principle, possess either of the two planar configurations shown in *Fig. 4*. However, the *cis* configuration is energetically less favourable (because carbon has a larger van der Waals radius than oxygen, and the two C_α groups get in each other's

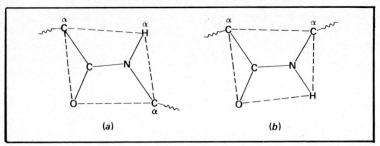

FIG. 4. The two alternative configurations of the planar peptide bond (*a*) *trans* and (*b*) *cis*.

way in *Fig. 4b*) and the *trans* configuration is almost invariably adopted in natural proteins.

A polypeptide chain, therefore, has rotational freedom only around the single bonds to each α-carbon atom, as indicated in *Fig. 5*. Even so, the number of different structures, or *conformations*,

FIG. 5. Rotation can still take place around the bonds to each α-carbon in a polypeptide chain.

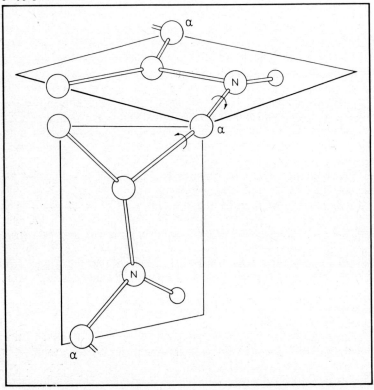

which the polypeptide backbone can adopt is very large. Not all conformations of the chain will have equal stability. For certain angles of alignment of successive peptide planes, there will be unfavourable van der Waals interactions as non-bonded groups are forced into contact. Conversely, other angles of alignment minimise such unfavourable interactions and, when adopted by a whole sequence of successive peptide units, the result is a strikingly regular folding of the chain. Particular stability is associated with folding schemes in which each peptide –NH– group is able to form a hydrogen bond to a peptide carbonyl group either within the same chain or in a separate chain.

In the right-handed α-helix, for example, the polypeptide chain is closely wrapped around the helical axis, with every peptide carbonyl group aligned parallel to this axis and within hydrogen bonding distance of the peptide –NH– group three residues further along the chain (*Fig. 6*). Since the amino acid side-chains project outwards they do not interfere, unless adjacent side-chains are identically-charged or highly-branched. Otherwise only a proline residue, with its special geometry, will disrupt the helical pattern.

One form of the so-called β-sheet structure is shown in *Fig. 7*. In contrast to the α-helix, each chain is fully extended, and forms strong hydrogen bonds to two closely neighbouring strands, one on either side, creating a corrugated sheet-like structure. Those parts of an enzyme molecule which show regular folding of the polypeptide chain, such as the α-helix or β-sheet, are referred to as *secondary structure*. Other aspects of the three-dimensional structure, including the disposition of side-chains, are usually referred to as *tertiary structure*. When an enzyme consists of two or more separately folded polypeptide chains that associate together by non-covalent bonds, it is said to possess *quaternary structure*.

In summary, the distinctive geometrical properties of the peptide bond place a fair amount of restriction on the three-dimensional structure of polypeptide chains. Certain conformations are particularly stabilised by the absence of unfavourable steric interactions and a correct geometry for the formation of strong hydrogen bonds. In enzymes, as in other globular proteins, regular folding patterns tend to play a supporting role, and an over-riding influence on the final structure is exerted by the amino acid side-chains.

The primary structure

The *in vivo* assembly of a polypeptide begins at the *N*-terminus, and the chain grows as amino acids are successively added. The information codified in the DNA sequence of the appropriate gene dictates which of the 20 common amino acids is inserted at each stage, and each copy of a given polypeptide therefore contains the

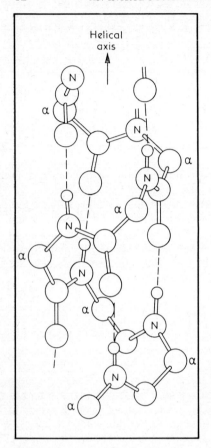

Helical
axis

FIG. 6.   The right-handed α-helix, in which hydrogen bonds are formed between peptide groups within the *same* chain.   The side-chains are not shown. (From Linus Pauling, *The nature of the chemical bond*. New York: Cornell University Press, and Oxford: OUP, 1960.)

same unique sequence of amino acid residues.   This unique sequence, referred to as the *primary structure*, confers a separate chemical identity on each enzyme protein.

The determination of primary structure

Because of its importance in determining the structure and reactivity of enzymes, there has been great interest in methods for determining primary structure.   For a large number of enzymes this complex task has already been accomplished: chicken lysozyme, for example, an enzyme in egg-white which catalyses the degradation of bacterial cell walls, is now known to have 129 amino acid residues in a single polypeptide chain.   The complete primary structure is shown in *Fig. 8*.   The abbreviations used for the amino acids are those of *Fig. 1*.

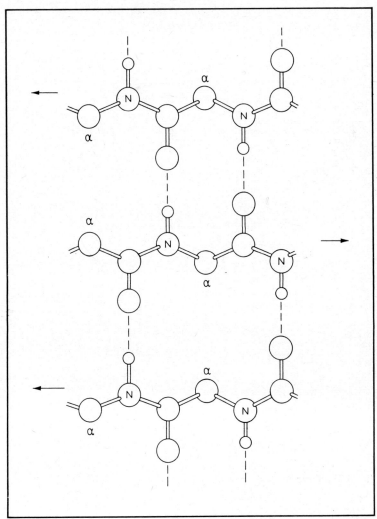

FIG. 7. One form of the β-sheet structure, in which hydrogen bonds link peptide groups in *different* chains. Side-chains are not shown.

The determination of primary structure requires several steps. Each polypeptide chain is split into small peptides by the action of chemical reagents or hydrolytic enzymes and the fragments are separated. The amino acid sequence of these short peptide fragments is then found by careful step-wise degradation of the chain, starting from the N-terminus, using an ingenious chemical method which allows the identification of each residue as it is

2

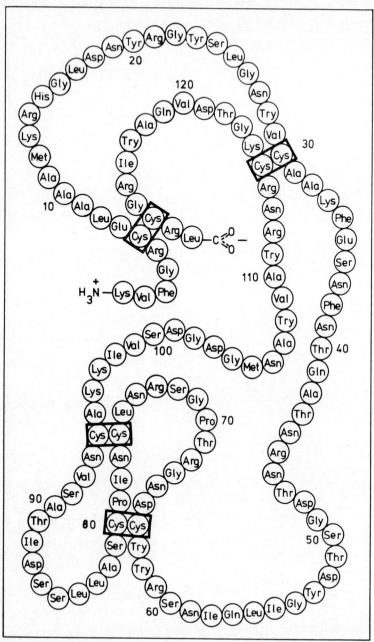

FIG. 8. The complete primary structure of chicken lysozyme. The abbreviations used for the amino acids are those of *Fig. 1*. (From R. E. Canfield and K. Liu, *J. biol. Chem.* ,1965 ,**240** ,2000.)

released. The next stage is to reconstruct the entire original sequence from the fragments. This is done by matching up sequences in peptides obtained from hydrolysis procedures of different specificity. Finally, the positions of any –S–S– bonds are determined. These steps are illustrated for lysozyme, in the following discussion.

To obtain the amino acid sequence of lysozyme in *Fig. 8*, the first step was to break the four –S–S– bonds by reduction with a thiol, and block the exposed cysteine thiol groups by alkylation (*Fig. 9*). This simplifies the subsequent procedures by preventing the formation of random –S–S– bonds during degradation of the molecule.

The initial fragmentation of the polypeptide was carried out using the digestive enzyme trypsin. This enzyme only attacks peptide bonds on the carboxyl side of the basic amino acid residues lysine and arginine. As may be verified in *Fig. 8*, this treatment should produce 18 different fragments including the C-terminal leucine as a free amino acid. In practice, five additional fragments were found, because certain reactive bonds were incompletely hydrolysed.

The short peptides were separated by ion-exchange chromatography and each one was degraded by the stepwise chemical technique, due to Edman, shown in *Fig. 10*.

Phenylisothiocyanate reacts (*a*) with the free α-amino group at the *N*-terminus of the peptide to form a phenylthiocarbamyl-peptide. On treatment with anhydrous acid (*b*) this cyclises and the peptide is expelled, shortened by one residue. The original *N*-terminal residue, with the side-chain R_1, is isolated after rearrangement (*c*), as its phenylthiohydantoin derivative. Chromatographic comparison with the reference substances prepared from the individual amino acids shows which of the 20 possible phenythiohydantoins

FIG. 9. The protection of cysteine thiol groups in lysozyme by (*a*) reduction followed by (*b*) alkylation.

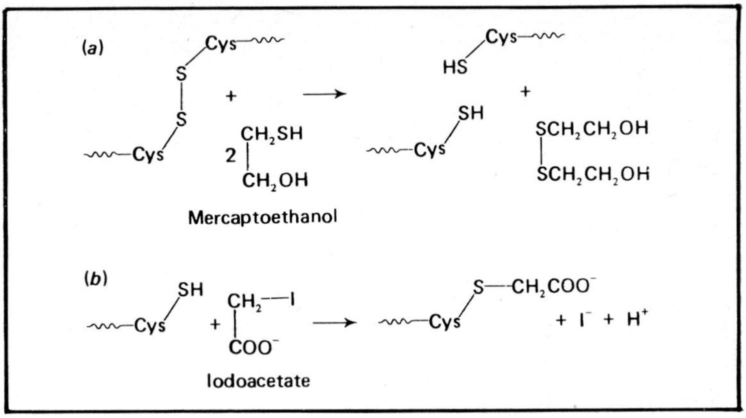

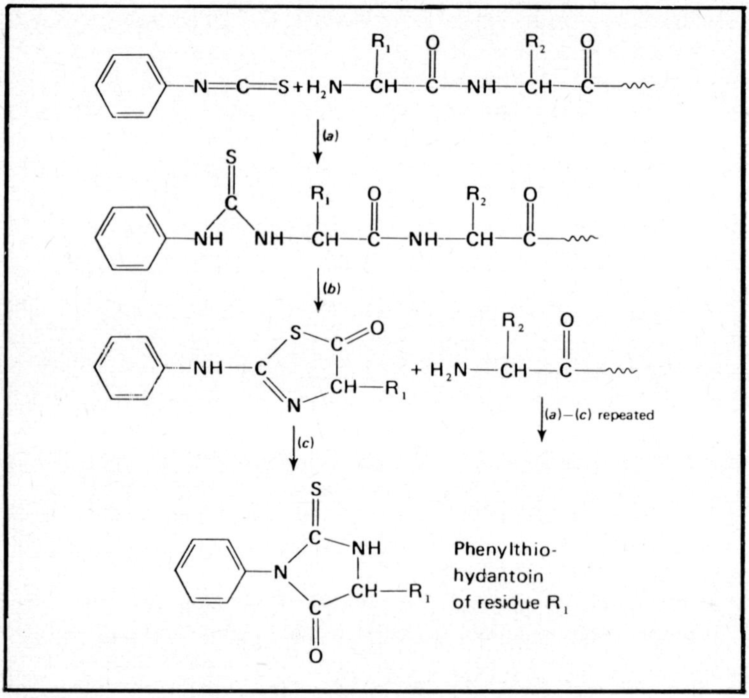

FIG. 10.   The Edman degradation procedure.

this is.   Alternatively, the shortened peptide may be re-submitted to amino acid analysis to see what is now missing.   The key feature of the Edman degradation is that it can be repeated on the shortened peptide, and in favourable cases 20 or more amino acids can be successively removed and identified.

When the fragments from the trypsin digest had been sequenced, the next problem was to arrange them in the correct order.   A second digestion of reduced, cysteine-protected lysozyme was carried out using chymotrypsin.   This enzyme prefers to attack peptide bonds on the carboxyl side of hydrophobic amino acid residues.   The resulting peptides were separated and sequenced, and provided much of the required overlap information.   For example, a 12-residue peptide with the sequence Ala-42 to Tyr-53 was isolated from the chymotrypsin digest (see Fig. 8).   This allowed the correct alignment of two of the original tryptic peptides, which had the sequences Phe-34 to Arg-45 and Asn-46 to Arg-61 respectively. Using this technique, and other chemical information, the entire 129-residue sequence could be deduced.

Finally, to locate the –S–S bridges, a sample of native lysozyme was treated with pepsin (yet another hydrolytic enzyme) at pH 2. Under these conditions the disulphide links are stable. After separation of the peptides using paper electrophoresis, partial sequence analysis of the cystine-containing fragments sufficed to locate their position in the molecule.

The chief aim of such chemical sequence studies has been to obtain a deeper insight into the link between the structure of a particular enzyme and its catalytic function. At the same time they have helped to pinpoint, in molecular terms, the differences between enzymes obtained from different organisms, which catalyse the same reaction. These comparative studies are of great current interest for theories of evolution at the molecular level. They have also been particularly valuable in focusing attention on certain parts of the sequence which appear to be identical — or nearly so — in the same enzyme from widely different organisms. Presumably these invariant portions of the molecule are vital for catalysis, or for the proper functioning of the enzyme *in vivo*.

It also appears that the same enzyme obtained from various tissues of a single organism will often differ to some extent in its primary structure, even if the difference is hard to detect in functional terms. Such closely-related species are called *iso-enzymes*. Recently, the recognition of certain human iso-enzymes has found important practical application in diagnostic medicine (*see* Chapter 5).

Three-dimensional structure of enzymes

The solubility of a globular protein in water can be diminished, by the addition of salts or organic solvent, to the point where it precipitates from solution. If this is done slowly, under carefully controlled conditions, the protein is often obtained in a crystalline form. Such crystals, in contrast to those of small organic molecules, usually contain a large volume fraction of liquid of crystallisation. The degree of long-range order is, though, sufficiently high that protein crystals will diffract x-rays to give strong, well-defined patterns.

This is exciting because it shows that (in the crystal at least) a globular protein has a well-defined structure, and that x-ray diffraction can be used to determine this structure. After many years of effort (the first diffraction pattern of a globular protein was obtained by Bernal and Crowfoot in 1934) the major experimental difficulties in the x-ray crystallography of proteins have now been overcome, and the crystal structures of a number of enzymes are now known to high resolution. In each case, the enzyme molecule appears to have a unique three-dimensional structure.

Some flexibility is obviously surrendered by the enzyme in the creation of a regular crystal lattice, but the conclusion seems warranted that, in solution as well, the catalytic activity is associated with an extremely well-defined structure, usually termed the *native* conformation. Recently-developed physical methods for analysing structural changes of enzymes in solution, such as nuclear magnetic resonance, largely confirm the results from the richly detailed, but static, x-ray picture. The additional flexibility revealed by the dynamic methods is included here in the term native conformation.

The idea of a unique active conformation also accounts for the observation that extremes of pH, or high temperatures, or the addition of urea or organic solvents, produce deep-seated conformational changes accompanied by complete loss of enzyme activity. These *denaturation* processes occur fairly readily, because most enzymes are maintained in their native conformation by a large number of relatively weak, non-covalent interactions.

The relationship between primary structure and the native conformation

Under favourable conditions of pH, temperature and concentration, denaturation may sometimes be reversed. For example, ribonuclease which has been reduced to break its four disulphide links, and then denatured using urea, becomes completely inactive. When the urea is removed by dialysis, the molecule is free to refold and, in the presence of molecular oxygen, to reform the four disulphide bonds. If this were to happen in a random fashion, then less than 1 per cent of the molecules would make all the four cross-links correctly, and reform the native conformation. Astonishingly, as indicated in *Fig. 11*, about 98 per cent of the molecules actually regain the native conformation and recover enzyme activity. The inescapable conclusion from this work is that the spontaneous folding of the chain causes it to adopt the conformation of *lowest free energy* in aqueous solution, and that this is identical with the native conformation.

This is experimental support for the hypothesis, originally stated by Crick in 1958, that the three-dimensional structure of an enzyme is simply a consequence of the particular sequence of amino acids (*ie* the primary structure). This hypothesis is now generally accepted, although much remains to be discovered about the folding process. For example, it appears that some enzymes fold up into the correct conformation extremely fast, so that only a fraction of the possible conformations can be 'inspected' by the chain before it settles on the one of highest stability. This would imply that the chain can direct its own folding, to allow it to attain the native conformation within a biologically useful time.

Maintenance of the native structure

The integrity of the native three-dimensional structure is maintained

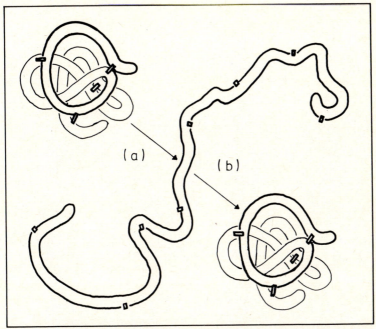

FIG. 11. The denaturation of ribonuclease. Reduction of the disulphide bonds and exposure to urea (*a*) leads to complete loss of enzyme activity. On removal of the denaturing agents and in the presence of molecular oxygen (*b*) almost all the enzyme molecules refold into the native conformation and regain activity. (From C. B. Anfinsen, Cold Spring Harbor Symposia on Quantitative Biology, Vol. 28, 1963.)

by weak, non-covalent interactions; these are hydrophobic interactions, hydrogen bonds, electrostatic interactions and London dispersion forces. In some proteins, covalent links are also important.

Hydrophobic interactions. It is common experience that hydrocarbons and other non-polar molecules are not miscible with water, so that, as an example, the free energy of transfer of benzene from the pure liquid into water is distinctly unfavourable, $\Delta G_t = +17$ kJ mol^{-1}. The free energy of transfer is made up of an enthalpic contribution, ΔH_t, and an entropic contribution ΔS_t

$$\Delta G_t = \Delta H_t - T\Delta S_t \qquad\qquad 4$$

where T is the absolute temperature. Since experiment shows that the enthalpy change is zero, the positive free energy of transfer is due entirely to an unfavourable change in the entropy of the system. A simple view of the origin of this effect is that the introduction of a hydrocarbon molecule into water leads to an increase in order among the water molecules closest to the hydrocarbon. The overall increase in order (decrease in entropy) is proportional to the surface area of the hydrocarbon. Conversely, the aggregation of two or more hydrophobic molecules in water is attended by a favourable entropy change. The aggregation diminishes the total surface area exposed to the solvent and allows the solvent molecules to become more disordered (*Fig. 12*). Hydrophobic side-chains are therefore expected to occupy positions in the core of the molecule, away from the solvent, and x-ray crystallography shows that this is largely true.

Hydrogen bonds. Hydrogen bonds between protein groups are not expected to possess much stability in water, which is a hydrogen-bonding solvent of high dielectric constant. Because of the hydrophobic interactions, however, a segment of polypeptide chain can find itself in less polar surroundings, and here the hydrogen bond makes an important contribution to defining the conformation. Generally, in the interior of a protein, the maximum number of hydrogen bonds are made. Secondary structure such as the α-helix or the β-sheet, with a regular pattern of main-chain hydrogen bonds, may be found either on the surface or in the core.

Electrostatic interactions. Usually the charged polar groups in proteins are on the surface and interact with solvent water. Electrostatic interactions between charged side-chains buried inside a protein are rare, although such an ion-pair should be rather stable.

London dispersion forces. The influence of attractive London dispersion forces is hard to estimate, but the interior of a protein is packed close enough that many van der Waals contacts are made, so they undoubtedly do help to maintain the native conformation once it is formed.

Covalent bonds. As already noted, the covalent disulphide bridges formed between cysteine residues in certain proteins are decisive in maintaining the conformation. In some enzymes, metal ions such as zinc or copper play a structural role, by bringing together as ligands two or more nucleophilic amino acid side-chains.

The active site
The distinctive kinetics of enzyme-catalysed reactions (*see* Chapter 3) strongly suggest that an enzyme functions by first binding its *substrates*, the compounds on which it acts, to form an enzyme-substrate complex. This is a specific recognition process, since many other molecules which may structurally resemble the substrates are bound less strongly, or not at all. Enzymes are typically very much larger than their substrates, so these observations lead naturally

to the idea of an *active site*, as that portion of the enzyme surface which is directly involved in binding and catalysis. As a rule, enzymes consisting of a single polypeptide chain possess only one

FIG. 12. The hydrophobic effect. Aggregation of benzene molecules in water reduces their exposed surface area, and hence the degree of order among solvent molecules.

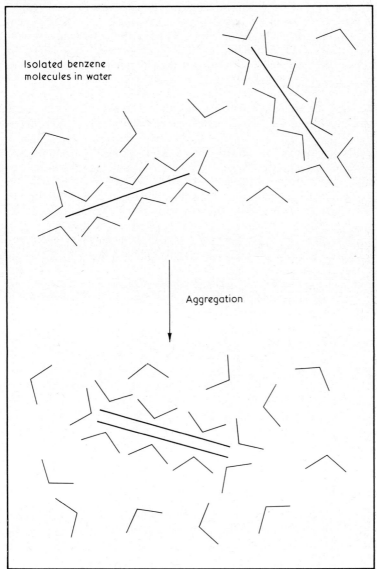

Isolated benzene
molecules in water

Aggregation

active site, while enzymes with quaternary structure may have several (usually one for each subunit).

Fischer, in 1895, was the first to advance the hypothesis that the enzyme and its substrate bind together in a complementary way, in the sense that a hole in a jig-saw puzzle is complementary to the piece that fits it.   The analogy he used was that of a lock and a key.   This accounts for the specificity of enzyme action because those molecules which are not sufficiently complementary to the active site will not be bound.   In addition, it explains the action of certain specific enzyme poisons, or *inhibitors*, for if these successfully bind at the active site, they will block the normal catalytic process.   The action of inhibitors is discussed further in Chapter 3.

X-ray crystallographic studies of enzyme structure support Fischer's hypothesis.   In the three-dimensional structure of many enzymes, the active site has been clearly discerned as a cleft or hollow in the surface of the enzyme.   Using scale models, it is possible to demonstrate a precise fit between a typical substrate and this portion of the enzyme surface.   The favourable interactions between enzyme and substrate are exactly analogous to those which maintain proteins in their native conformation: hydrophobic and ionic interactions, hydrogen bonds, and short-range van der Waals forces.

The weakness of the Fischer hypothesis lies in the assumption that the enzyme is completely rigid.   As already noted, it is likely that enzymes in solution possess a measure of flexibility and it has frequently been possible to detect — using a range of spectroscopic methods — changes in enzyme conformation accompanying the binding of substrates.   An extension of the Fischer hypothesis has been proposed by Koshland, which takes into account such conformational changes, and links them to the need for an enzyme to assemble, at the active site, several individual amino acid side-chains for catalysis to occur.   Koshland's 'induced fit' hypothesis is illustrated schematically in *Fig. 13*.   In order that amino acid residues 1 and 2 can both interact with the portion X in the substrate, advantage is taken of the specific binding between the complementary regions A and A*, B and B* on the substrate and the protein.   The energy needed to distort the enzyme conformation is provided by the favourable free energy change for substrate binding.

This hypothesis provides an explanation for some observations of enzyme specificity which do not fit the Fischer model.   For example, consider the replacement of the substrate in *Fig. 13* by a compound with the same features, but lacking the region labelled B.   Such a compound would be able to bind to the enzyme, perhaps quite strongly, but would not be able to induce the conformational change which leads to reaction.   Although the analogue appears to contain the critical feature of the substrate, the region X, it will fail to react,

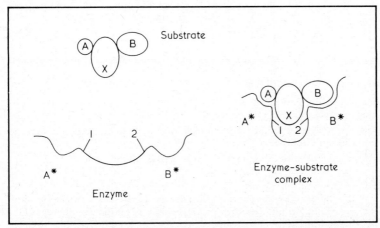

FIG. 13. The 'induced fit' model, in which specific interactions between A and A*, B and B*, induce a conformational change in the protein so that both amino acid side-chains 1 and 2 interact with the substate at X. (From W. P. Jencks, *Catalysis in chemistry and enzymology*. New York: McGraw-Hill, 1969.)

despite being small enough to penetrate easily into the same active site.

The role of cofactors

For a large number of enzymes, the presence of relatively small organic molecules (relative molecular mass less than 1000 or so) is required for activity. These cofactors are the active forms of the vitamins and, although rather few in number, they show great diversity in chemical structure. A distinction is usually drawn between *prosthetic groups*, which remain tightly bound to an enzyme active site and are continuously regenerated during the enzyme-catalysed reaction, and *coenzymes*, which dissociate more or less freely from the enzyme and which are required in stoichiometric amounts. However, the role of the cofactor is the same in each case; amino acid side-chains only provide a range of Brønsted acids and bases, and cofactors enable a much greater variety of transformations to be catalysed. For example, the coenzyme nicotinamide adenine dinucleotide (NAD), discussed in Chapter 4, enables the enzyme alcohol dehydrogenase to carry out an oxidation–reduction process. Metal ions, such as zinc and magnesium, are also frequently involved as cofactors in enzyme-catalysed reactions.

2. General Principles of Enzyme Catalysis

General considerations

Any reaction susceptible to catalysis must be both chemically and thermodynamically possible. This is easy to see for fast reactions such as the interconversion of carbonic acid and dissolved carbon dioxide in blood

$$H_2CO_3 \rightleftharpoons H_2O + CO_2$$

which reaches equilibrium within minutes, at pH 7 and 37 °C, even in the absence of the enzyme carbonic anhydrase. It is less obvious for reactions which are immeasurably slow unless the appropriate enzyme is added. The enzyme, like any other catalyst, in each case simply speeds up the rate of approach of the system towards equilibrium.

The rate of any reaction is determined by the free energy of activation $\Delta G^{\neq}$, which is the difference in free energy between the starting materials and the transition state which leads to product. This is shown schematically by the energy profile in *Fig. 14*. In such diagrams the overall free energy of the system is plotted against the reaction coordinate which represents the degree of progress towards product formation. Only the relative energy levels of maxima or minima in the profile have physical significance. In formal terms, the enzyme speeds up the rate of product formation by lowering the energy barrier to reaction, $\Delta G^{\neq}$. In the reverse direction, the corresponding barrier is $\Delta G^{\neq} + \Delta G_0$, where ΔG_0 is the net free energy change for the reaction, so that both forward and reverse reactions will benefit from a lowering of $\Delta G^{\neq}$.

It is a valuable kinetic principle that the pathway favoured for the forward reaction is also favoured for the reverse reaction. A direct consequence of this principle of microscopic reversibility is that every product of an enzyme-catalysed reaction will itself be a substrate for the reverse reaction. In practice, as will be discussed later in this Chapter, it is misleading to think of the catalysed reaction as 'the same reaction speeded up'. It is more appropriate to say that the enzyme provides a new alternative reaction pathway, for which the free energy of activation is lower.

As true catalysts, enzymes are effective at concentrations very much lower than those of the other reaction components. Under such conditions, the concentrations of free substrates and products at equilibrium will clearly be exactly the same as in the absence of the catalyst. It should be remembered, though, that *in vivo* many intracellular enzymes are present in high (often localised) concentra-

24

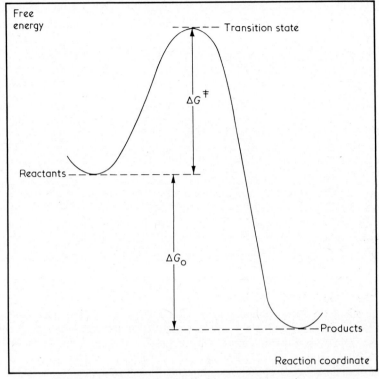

Fig. 14.   Free energy profile for a one-step reaction.

tions relative to their substrates, and such substrates are likely to exist largely in an enzyme-bound form.

Another fundamental property of catalysts is that they themselves suffer no net change as a result of catalysis.   As each molecule of product departs from the active site, the enzyme is returned to exactly the state which originally combined with the substrate.   This spontaneous regeneration, or turnover, means that the same catalytic site may be re-used indefinitely.   In practice, the 'life' of an enzyme catalyst is limited by factors such as an intrinsic instability under the reaction conditions, or the presence of impurities which progressively 'poison' the catalytic sites.

Homogenous catalysis

Two types of catalysis in aqueous solutions are particularly relevant to enzyme function.   These are covalent catalysis and general acid–general base catalysis.

Covalent catalysis

Many enzymes react chemically with their substrates to form covalently-bonded intermediates. A simple model for such reactions is provided by the effect of iodide ion on the hydrolysis of methyl bromide. The hydrolysis involves nucleophilic bimolecular substitution (S_N2) of bromide ion by water, followed by a fast proton transfer to water.

$$H_2O + CH_3-Br \xrightarrow{\text{slow}} H_2O^+-CH_3 + Br^-$$

$$H_2O^+-CH_3 + H_2O \xrightarrow{\text{fast}} HO-CH_3 + H_3O^+$$

In the presence of small amounts of iodide ion, the hydrolysis is greatly accelerated. Catalysis is brought about by the nucleophilic attack of iodide ion on methyl bromide, forming a covalent intermediate, methyl iodide. This subsequently hydrolyses in a second nucleophilic substitution reaction.

$$I^- + CH_3-Br \longrightarrow I-CH_3 + Br^-$$

$$I-CH_3 + H_2O \longrightarrow CH_3-OH + H^+ + I^-$$

This reaction illustrates the essential requirements for effective covalent catalysis. The covalent intermediate is formed faster, and breaks down faster, than the direct substitution can occur (since iodide ion is both a better nucleophile than water, and a better 'leaving group' than bromide ion). Secondly, methyl iodide is thermodynamically less stable than the final products of hydrolysis, which prevents any tendency for it to accumulate.

General acid–general base catalysis

Chemical reactions of biological interest are often subject to acid or base catalysis, and sometimes both. Where the rate enhancement is due to the presence of free hydronium ions (H_3O^+) or hydroxide ions (OH^-) this is termed *specific* acid or base catalysis respectively. If reactions are also promoted by other proton donors and acceptors, this is *general* acid or base catalysis. Since enzymes are designed to function at or near neutral pH values, and they contain a range of Brønsted acids and bases, general acid–base catalysis is obviously attractive as a possible contributory factor in enzyme action.

General catalysis frequently benefits reactions involving slow proton transfers, such as the enolisation of acetone (propanone) (*Fig. 15a*). The two alternative ways in which a general acid HA might affect this reaction are shown in *Fig. 15b* and *15c*. In the first, the transition state, with a developing negative charge on the carbonyl oxygen, is stabilised by partial proton transfer from HA. In the other possible mechanism (which turns out to be correct) the general

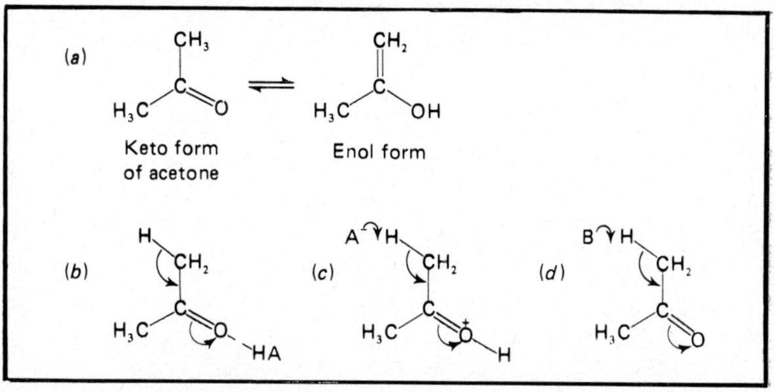

FIG. 15. (*a*) The interconversion of the keto and enol forms of acetone. (*b*) and (*c*) two alternative mechanisms for catalysis of enolisation by a general acid HA. (*d*) A mechanism for catalysis by a general base B. The curved arrows represent the movement of electron pairs in the course of the reaction.

acid transfers a proton rapidly and completely to acetone, and then its conjugate base A^- attacks the conjugate acid of acetone. These two mechanisms are kinetically indistinguishable from one another, because in each case the rate of reaction will show the same dependence on the concentration of HA. This mechanistic ambiguity is an ever-present problem in the study of general catalysis and additional evidence is needed to resolve it in each case. In the same way, two kinetically equivalent mechanisms could be written for the catalysis of acetone enolisation by a general base B, although in fact the correct one is indicated in *Fig. 15d*.

Another class of reactions susceptible to general acid and base catalysis is exemplified by the nucleophilic addition of water to a carbonyl group (*Fig. 16a*). Here, the rate of formation of the carbon–oxygen bond, rather than proton transfer, is the slow step. The intervention of a general base B (*Fig. 16b*) or of a general acid HA (*Fig. 16c*) is beneficial because in each case an alternative reaction pathway is provided, which does not involve a relatively unstable dipolar intermediate. Again, it is worth noting that another, kinetically equivalent, mechanism may be formulated for both general acid and general base catalysis. Fortunately, in the highly organised active site of an enzyme, it is usually possible to determine which mechanism is actually operating.

Enzyme catalysis

In the absence of any satisfying general theory of enzyme catalysis, the best that can be done is to identify various factors which are likely to contribute to the observed rate enhancement. Once the enzyme-substrate complex is formed, the mechanisms discussed

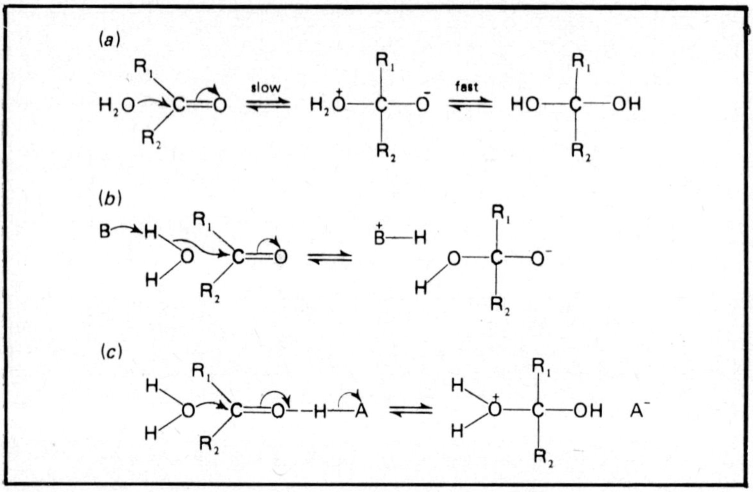

FIG. 16. (*a*) The addition of water to a carbonyl group, (*b*) catalysis of the reaction by a general base B and (*c*) catalysis by a general acid HA.

above under homogeneous catalysis may be relevant for many enzymes, but other factors, including approximation, orientation and strain effects, are important both for the formation of the enzyme-substrate complex and for the subsequent steps of catalysis.

Approximation, orientation and strain

These are usefully taken together, because their specific contributions are not easily separable. Approximation refers to the 'collection' together of the reactants at the active site. To see how this might have kinetic consequences consider the free energy of activation for an uncatalysed bimolecular reaction in solution, $\Delta G^{\ddagger}$. This may be expressed as

$$\Delta G^{\ddagger} = \Delta H^{\ddagger} - T\Delta S^{\ddagger} \qquad 5$$

where $\Delta H^{\ddagger}$ and $\Delta S^{\ddagger}$ are the enthalpy and entropy of activation respectively, and T is the absolute temperature. For a bimolecular process, $\Delta S^{\ddagger}$ will generally be rather unfavourable (negative) because some of the translational degrees of freedom in the reactants must be lost in reaching the transition state. When the reactants are brought together at the active site of an enzyme to form an enzyme-substrate complex (ES), a great deal of this entropy is lost.

The further entropy loss required in going from ES to the transition state is therefore comparatively small, and this is reflected in a lower free energy of activation for this step. The more favourable entropy of activation for the reaction within the enzyme-substrate complex therefore results in a lower Gibbs free energy of activation for the catalysed reaction ($\Delta G_{cat}^{\ddagger}$). This approximation effect is shown schematically in the free energy diagram of *Fig. 17*.

When the enzyme-substrate complex is formed, the reactants also lose part of their rotational freedom and this can provide a further entropic contribution to catalysis. Most reactions with an appreciable activation energy demand a well-defined spatial relationship between the reacting species at the transition state, and if substrate binding removes rotational degrees of freedom so that this 'correct' orientation is favoured, this will also contribute to the rate enhancement.

Extensive use has been made of intramolecular reactions as models for the effects of approximation and orientation. Here the reaction partners are incorporated into the same molecule so that the reaction becomes, like the rate-determining step in enzyme catalysis, a unimolecular process. Compare the two esterification processes in *Fig. 18*. The rate-constants for the unimolecular reaction (*b*) and the bimolecular reaction (*a*) have different dimensions, so that they cannot be directly compared. One approach is simply to divide the first-order rate constant k_1, by the second order rate constant k_2, and consider the concentration obtained (79 M in the example given) as

FIG. 17. The approximation effect. Free energy profiles for (– – – –) an enzyme-catalysed reaction and (————) for the same reaction in the absence of a catalyst.

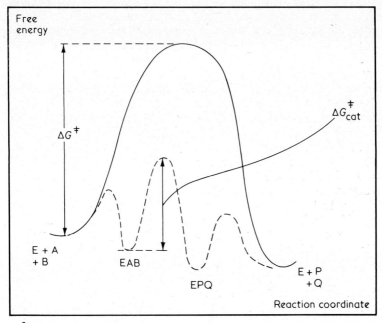

3

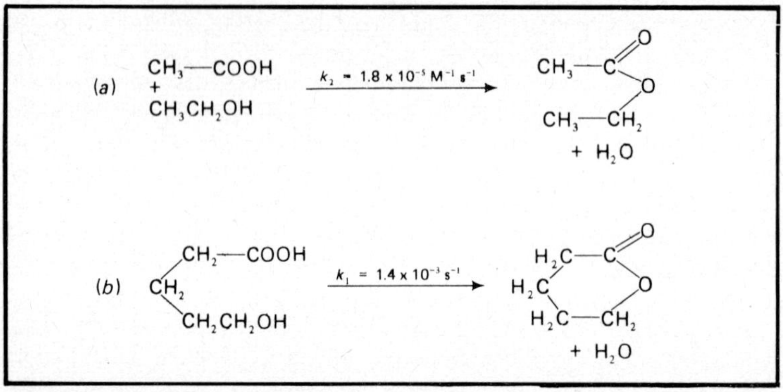

FIG. 18. Comparison of the rate of an intermolecular reaction (a) with that of its intramolecular counterpart (b). This provides a chemical model for the effects of approximation and orientation.

an 'effective molarity' of each reacting species in the intramolecular reaction. Unfortunately, the speed of intramolecular reactions compared to their intermolecular counterparts, which otherwise proceed by the same mechanism, is often too great to be attributed to a simple juxtaposition of the reacting species. The concentration in water of water itself is only 55 M, so the physical meaning of any 'effective molarity' greater than this value is unclear.

The question then becomes, whether the loss of translational and rotational entropy entailed in fixing the reactant species into the same molecule by a covalent bond is sufficiently large quantitatively to account for the observed rate enhancements. Only approximate calculations can be made, but it has been suggested that rate enhancements of up to $10^8 M$ can be accounted for on this basis. This view of enzyme catalysis implies that the enzyme-substrate complex, far from being a loose 'encounter complex', is a much more intimate state, in which the substrate fits very snugly into the active site, many van der Waals contacts are made, and the mobility of the substrate, with respect to the enzyme's catalytic groups, is restricted. This also accords satisfyingly with the results of high-resolution x-ray structural determinations which clearly demonstrate the excellent fit between an enzyme and its specific substrate. This use of intrinsic binding energy to 'freeze' reactants at the active site of an enzyme also provides an interesting explanation for the apparent complexity of many enzyme cofactors (*see*, for example, *Fig. 35*). The complex structure obviously aids the correct recognition of these compounds by the enzyme, but, in addition, the extra binding energy may be used to 'anchor' the molecule securely at the active site, maximising the loss of translational and rotational entropy in the binding step.

A complete analysis must also take account of strain effects, since model studies show that intramolecular reactions are particularly sensitive to the degree of constraint in the reactants. If reactants are under some form of geometric or electronic strain, and this is wholly or partially relieved in the transition state for the reaction, then the free energy of activation will clearly be lower than for an analogous strain-free system. A general mechanism by which an enzyme could exploit strain effects is illustrated in *Fig. 19*. It shows two hypothetical free energy profiles, one for a substrate binding tightly without strain, and a second for the same substrate, in which much of the binding energy is used to force the enzyme-substrate complex into a strained condition, the geometry of which more nearly resembles the transition state.

As first pointed out by Hammond, when adjacent species on a reaction coordinate are similar in structure, they will also have a similar free energy content. When binding energy is used to force the enzyme-substrate complex to adopt a structure which more closely resembles that of the transition state, then, according to Hammond's postulate, the free energy difference between the two (*ie* the free energy of activation) will also be decreased. This could also be described as an orientation effect, in a case where the binding energy is used to constrain an enzyme-substrate complex in a parti-

FIG. 19. A hypothetical free energy diagram showing the use of binding energy to force the ES complex along the reaction coordinate (———), lowering the free energy of activation, rather than to optimise the strength of substrate binding (– – – –).

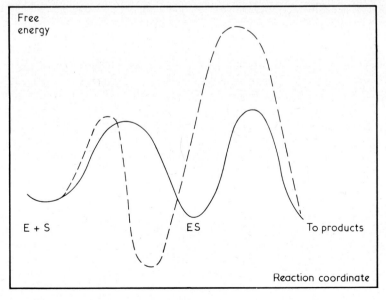

cular orientation which makes it more nearly resemble the geometry of the transition state. In any event, the effect is to reduce the free energy of activation by destabilising the ground state (the ES complex) with respect to the transition state.

Concerted catalysis

An enzyme which requires no metal ions or cofactors for activity must clearly rely on the chemical reactivity of its amino acid side-chains, and in particular on their acid–base properties. As already mentioned, a Brønsted base can function in catalysis either as a general base (a proton acceptor) or as a nucleophile. A Brønsted acid can act as a source of hydronium ion (specific acid catalysis) or, directly, as a general acid catalyst. An attractive possibility is that part of the very large rate enhancement in enzyme catalysis is due to the cooperative action of two 'components', for example a nucleophile and a general acid, or a general base and an electrophilic metal ion.

Evidence for this concerted or 'push–pull' catalysis has been sought for model reactions in aqueous solution. The enolisation of acetone, for example, is accelerated in the presence of both acetic (ethanoic) acid (a general acid) and acetate ion (a general base). A possible concerted mechanism is shown in *Fig. 20*.

In practice, this pathway is found not to contribute much to the observed catalysis. Evidently the loss in translational and rotational entropy involved in bringing together acetone and two molecules of catalyst outweighs any possible kinetic advantage. The transition state is simply too improbable.

In an enzyme active site, on the other hand, the catalytic residues are already fixed in position, so that concerted catalysis is much more likely. The hydrophobic environment provided by most enzyme active sites also encourages such a mechanism, because favourable ionic interactions are promoted and because water is largely excluded as a proton donor or acceptor. Kinetic evidence for 'concertedness' is difficult to obtain, but in several cases where the detailed structure of the active site is known, it is intriguing to discover that proton

FIG. 20. A possible mechanism for concerted general acid–base catalysis of the enolisation of acetone by acetic acid and acetate ion.

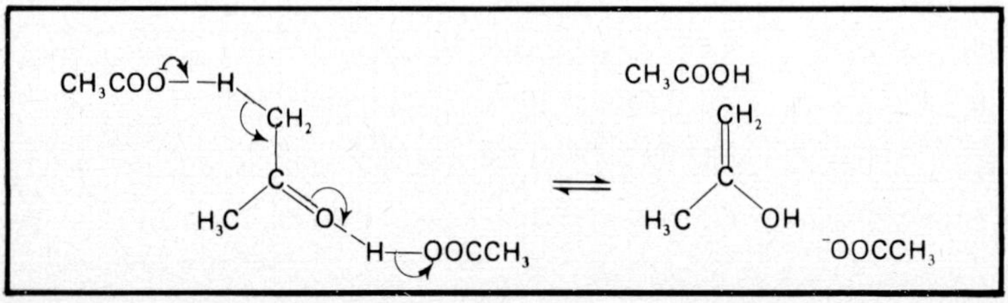

donors and acceptors are in precisely the correct positions for this mechanism to operate.

The role of intermediates

As discussed earlier in this Chapter, many reactions in homogeneous solution benefit from covalent catalysis, in which the catalyst offers a favourable alternative reaction pathway involving an intermediate.

Covalent catalysis is particularly common among enzymes which catalyse group transfer reactions, or which depend on cofactors or coenzymes for activity. In the latter case the intermediate is formed from the substrate and the enzyme-bound cofactor.

In a direct group transfer reaction catalysed by an enzyme

$$AX + B \rightleftharpoons A + BX$$

much of the translational and rotational entropy of both AX and B must be lost on binding to the enzyme and reaching the transition state for the reaction. If, though, the reaction proceeds in two consecutive steps

$$AX + E \rightleftharpoons A + EX$$
$$EX + B \rightleftharpoons E + BX$$

the transition state for formation of EX can be reached without losing any of the entropy of B, and B is only immobilised in a second, separate transition state. The entropy of activation of each step will, therefore, be considerably more favourable than that for the direct transfer and this provides a distinct catalytic advantage for the two-step pathway.

Enzyme specificity

Specificity, which is widely regarded as the hallmark of enzyme catalysis, is manifested in several important ways. Reaction specificity, for example, refers to the type of chemical reaction that is catalysed, and provides the basis for a systematic enzyme classification into six major groups. Thus the *oxidoreductases* catalyse oxidation–reduction reactions; the *transferases* catalyse the transfer of atoms or groups between different molecules; hydrolytic reactions are catalysed by *hydrolases*, and additions to double bonds by *lyases*; *isomerases* catalyse geometrical or structural rearrangements within the same molecule and *ligases* catalyse certain energy-requiring synthetic reactions.

The substrate specificity of an enzyme defines its function much more closely. In general, enzymes act on rather a limited number of substrates and often only one reaction is catalysed with any degree of efficiency. The unique three-dimensional structure of an enzyme clearly allows it to recognise quite small chemical differences between potential substrates. Even enzymes that accept a large number of substrates may show a clear specificity in their site of action. The hydrolase trypsin, for example, attacks most natural polypeptides, but the hydrolysis always occurs at a peptide bond on the carboxyl side of either a lysine or an arginine residue. Recent x-ray crystallographic studies have shown that the folding of the enzyme chain creates a 'pocket' at the active site, at the bottom of which is a negatively-charged carboxyl group. This site exactly accommodates the positively-charged side chain of lysine and arginine, and rejects the other 18 amino acids. Such binding effectively helps to anchor the peptide in a favourable conformation for the subsequent bond cleavage.

Enzyme stereospecificity

Every enzyme is chiral, because it is constructed from L-amino acids which are themselves chiral, and because of chiral features of the secondary and tertiary structure. This is of great importance for enzyme stereospecificity, because a chiral reagent is able to discriminate between any two stereoisomers of a compound, whereas an achiral reagent can only discriminate between stereoisomers which have different energy contents and, therefore, different physical properties. As already noted in Chapter 1 (*see Fig. 3*), almost all the α-amino acids can exist in either of two mirror-image forms (enantiomers) which have identical energy contents. Generally only one enantiomer is accepted as a substrate by a given enzyme. For example, L-amino acids such as L-alanine are rapidly oxidised in the presence of L-amino acid oxidase, but the corresponding D-enantiomers are recovered unchanged (*Fig. 21*).

Its chirality also enables an enzyme to discriminate between

FIG. 21. The oxidation of L-alanine by L-amino acid oxidase. D-amino acids are not attacked.

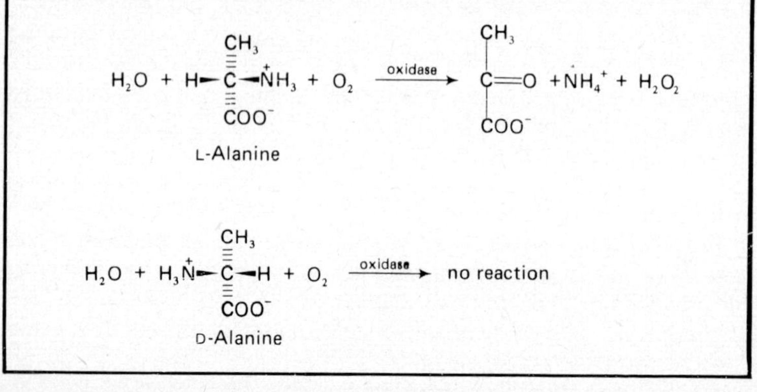

chemically identical groups within the same molecule which are sterically non-equivalent. For example, glycerol (propane-1,2,3-triol) contains two hydroxymethene groups which occupy mirror-image positions in the molecule:

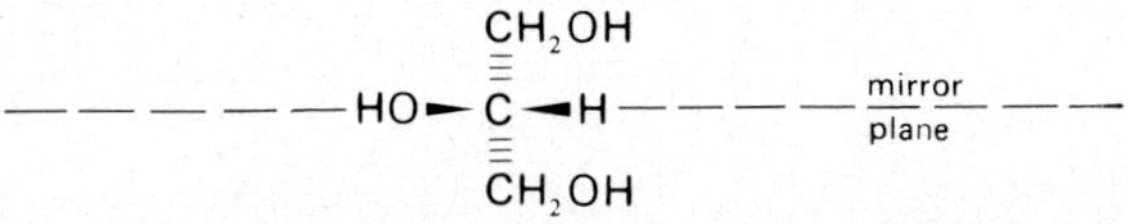

$$CH_2OH$$
$$------HO-C-H------ \quad \frac{mirror}{plane} ---$$
$$CH_2OH$$

An enzyme can readily differentiate between these two hydroxymethene groups; thus, in the phosphorylation of glycerol by adenosine triphosphate (usually abbreviated ATP) in the presence of glycerol kinase, the products are adenosine diphosphate (ADP) and only one enantiomer of glycerol phosphate, as indicated in *Fig. 22*.

In studying this aspect of enzyme stereospecificity, it is essential for the investigator to be able to put his own mark specifically on one of the two chemically identical atoms or groups, so that he can always tell them apart. This is done using isotopes. The carbon isotope ^{14}C and the hydrogen isotope tritium ^{3}H are often chosen because their radioactivity makes them easily detectable. Isotopic substitution introduces predictable effects due to the different atomic mass, but otherwise the chemical reactivity is unchanged. For example, a specific ^{14}C label in glycerol (denoted by an asterisk in *Fig. 22*) acts as a convenient tracer for the fate of that individual atom in the glycerol kinase reaction.

When paired, chemically identical groups or atoms in a substrate

FIG. 22. The phosphorylation of glycerol by ATP catalysed by glycerol kinase. The phosphoryl group (shown here in the fully ionised form) becomes attached specifically to the asterisked hydroxyl group.

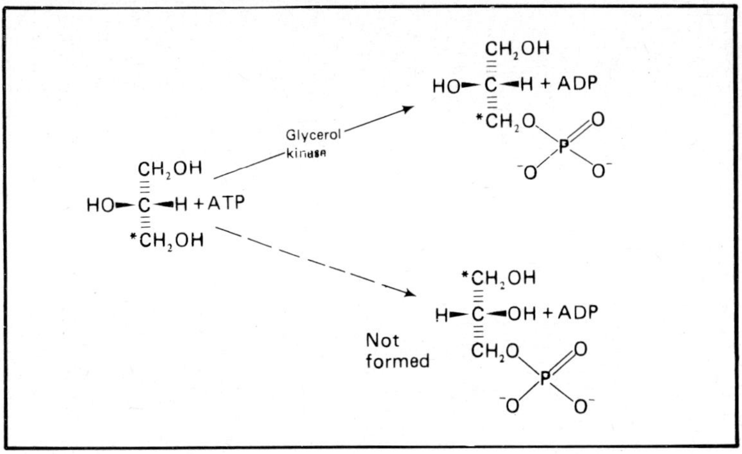

are sterically equivalent, even an enzyme cannot differentiate between them. Ethylene glycol, for example, has two hydroxymethene groups that are entirely equivalent. A simple rotation of the molecule through 180° generates an arrangement which is indistinguishable from the original:

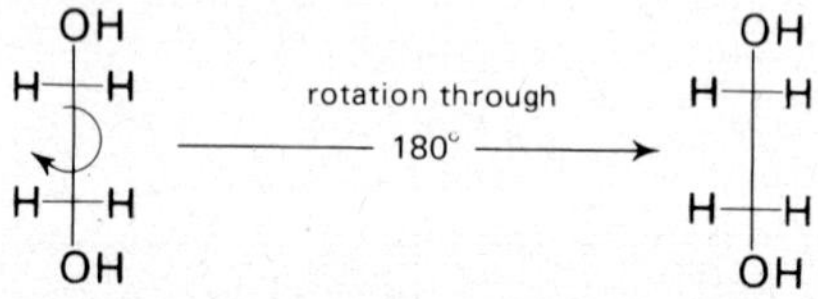

The two hydroxymethene groups are, therefore, attacked indiscriminately by suitable enzymes.

The earliest attempt to explain how an enzyme differentiates between chemically identical groups was made by A. G. Ogston in 1948. He pointed out that if a molecule was bound to an enzyme through at least three specific points of contact, paired groups would be placed in non-equivalent positions. This is shown schematically in *Fig. 23*. This idea was crucial to the development of enzyme stereochemistry, and it places useful emphasis on the importance of a complementary fit between enzyme and substrate. Unfortunately, it has often been misinterpreted to mean that the enzyme somehow induces asymmetry in its substrate. An enzyme only discriminates between molecules or groups that are inherently non-equivalent.

FIG. 23. The Ogston hypothesis. The existence of three specific points of contact (AA', BB' and CC') ensures that the hydroxymethene groups of glycerol occupy different parts of the enzyme active site.

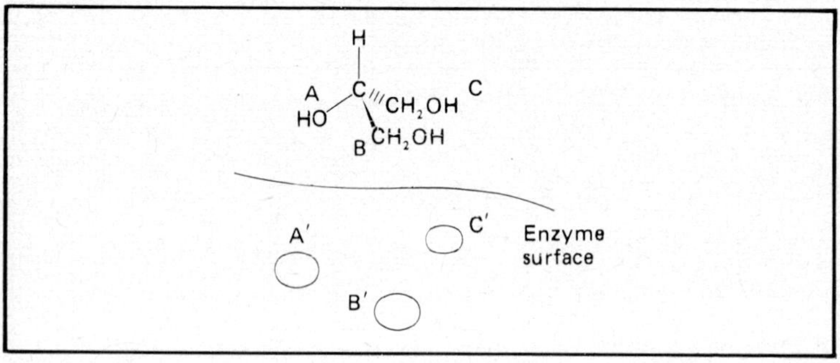

3. Enzyme Kinetics

Introduction

The kinetics of enzyme-catalysed reactions are routinely studied using extremely low enzyme concentrations. This is experimentally convenient, because reactions are then completed within minutes or hours rather than within fractions of a second, and their rates can be followed by conventional techniques. Initially, the rate is linear (*Fig. 24*) but in practice it often decreases at a later stage (because of progressive enzyme denaturation, for example).

As expected for any catalyst, the initial rate of an enzyme-catalysed reaction is directly proportional to the concentrations of enzyme, if all other parameters are kept constant. When the enzyme concentration is held fixed, and the initial substrate concentration [s] is increased, the initial rate of the reaction v_0 varies as shown in *Fig. 25*.

At the beginning of this century, Henri, Michaelis and others recognised that this distinctive kinetic behaviour could be rationalised

FIG. 24. Typical progress curve of an enzyme-catalysed reaction.

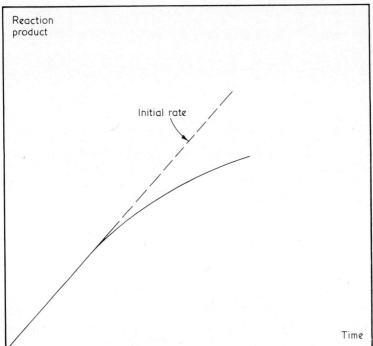

37

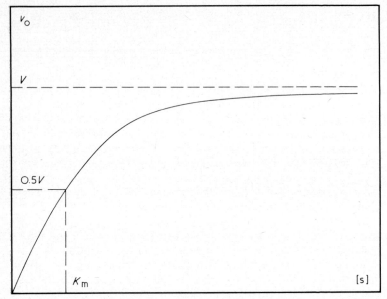

FIG. 25. Plot of initial velocity v_0 of an enzyme-catalysed reaction as a function of the substrate concentration [s].

by assuming that catalysis occurs through the intermediate formation of an enzyme-substrate complex:

$$E + S \rightleftharpoons ES \rightarrow E + P$$

On the basis of this scheme they derived an equation of the form

$$v = \frac{V[s]}{K_m + [s]} \qquad\qquad 6$$

which correctly describes the curve of *Fig. 25*. V and K_m are characteristic of each enzyme-substrate combination. The parameter V represents the maximum velocity obtained when the enzyme is completely saturated with substrate. Thus, when $[s] \gg K_m$, $v_0 = V$.

In comparing the catalytic effectiveness of an enzyme with different substrates it is often useful to know the maximum number of substrate molecules acted upon by each active site in unit time. This is the turnover number, k_{cat}, which has the dimensions of a first order rate constant and is defined by

$$k_{cat} = \frac{V}{n[e]_0}$$

where $[e]_0$ is the enzyme concentration and n the number of active sites on each enzyme molecule. The parameter K_m, usually referred to as the Michaelis constant, is defined by equation 6 as the value of [s] at which $v_0 = 0.5\ V$.

The importance of equation 6, the Michaelis–Menten equation, is that it adequately describes the kinetic behaviour of a very large number of enzymes. Where the reaction involves two or more co-substrates, a separate equation of this form still describes the dependence of v_0 on each of the substrate concentrations, if the concentrations of the others are kept constant.

The steady state

Michaelis and Menten assumed, in deriving their equation, that the enzyme-substrate complex and the starting materials are always at equilibrium, and that the breakdown of the enzyme-substrate complex into products is much the slowest step. Briggs and Haldane were later able to derive an equation of exactly the same form using an alternative and much more reasonable assumption. They used the same simple kinetic scheme:

$$E + S \underset{k_2}{\overset{k_1}{\rightleftharpoons}} ES \xrightarrow{k_3} E + P$$

where k_1, k_2 and k_3 are the rate constants for the individual steps. Briggs and Haldane assumed that after a very short reaction time — a matter of milliseconds — the concentration of enzyme-substrate complex has built up to a steady value, where its rate of formation is exactly balanced by its rate of decomposition. The concentration of the enzyme-substrate complex thus remains constant at a small, steady-state value during the period when initial velocity measurements are made. This assumption of a steady state is generally valid under the usual conditions of such measurement where [s], the substrate concentration, greatly exceeds the total enzyme concentration $[e]_0$. The rate of formation of the ES complex is given by

$$\frac{d[es]}{dt} = k_1[e][s] \qquad 7$$

and its rate of decomposition by

$$-\frac{d[es]}{dt} = (k_2 + k_3)[es] \qquad 8$$

where [es] is the concentration of enzyme-substrate complex and [e] the concentration of free enzyme.

In the steady state

$$k_1[e][s] = (k_2 + k_3)[es] \qquad 9$$

so that

$$[es] = \frac{k_1[e][s]}{(k_2 + k_3)} \qquad 10$$

Since

$$[e] = [e]_0 - [es] \qquad 11$$

10 becomes, after rearrangement,

$$[es] = \frac{[e]_0[s]}{[s] + \dfrac{(k_2 + k_3)}{k_1}} \qquad 12$$

The initial velocity v_0 measures the rate of breakdown of ES into products:

$$v_0 = k_3[es] \qquad 13$$

Therefore, from 12 and 13

$$v_0 = \frac{k_3[e]_0[s]}{[s] + \dfrac{(k_2 + k_3)}{k_1}} \qquad 14$$

This has exactly the form of equation 6, the original Michaelis–Menten equation, where

$$V = k_3[e]_0 \quad \text{and} \quad K_m = \frac{k_2 + k_3}{k_1}$$

The steady state treatment thus gives physical significance to the parameters in the Michaelis–Menten equation, in terms of the rate constants for individual steps. This kinetic treatment can readily be extended to more complex systems involving two or more co-substrates and several distinct enzyme-bound intermediates. The kinetic effects of the reverse reaction, of inhibitors or of changes in pH can all be satisfactorily treated within the same framework. In general, the expressions for K_m and V will contain a complex collection of individual rate constants, but frequently other information about the mechanism allows these expressions to be simplified.

Rapid reaction techniques

Since most enzyme-catalysed reactions reach a steady state within a matter of milliseconds after the mixing of enzyme and substrate, the term 'initial' velocity usually refers to this phase of the reaction. Using specially designed equipment, it is possible to study the events which lead up to the steady state. In the so-called stopped-flow method, substrate and enzyme are delivered automatically from syringes, mixed and then forced into an observation cell. When the cell is full the flow is stopped automatically, and the subsequent chemical changes in the mixture are monitored spectrophoto-metrically. Thus the complete time course of the reaction may be followed, after the one or two milliseconds required for mixing. This usually involves both the changes that occur as the steady state is reached, and the measurement of the steady state rate as well.

Mathematical treatment of the pre-steady state phase is not trivial, but such methods provide kinetic information which usefully supplements that obtained from steady state measurements. In particular, stopped-flow techniques may allow direct observation and identification of enzyme-bound intermediates which are present in vanishingly small concentrations during the steady state.

A completely different approach to the study of fast reactions is embodied in what are termed relaxation methods. Here, a system at chemical equilibrium is perturbed by an 'instantaneous' change in an intensive thermodynamic property, such as temperature or pressure. This changes the equilibrium constant, and the concentrations of all reaction components must then adjust themselves to their new equilibrium values. Where, as in enzyme-catalysed reactions, several species are present at equilibrium, the rate of chemical relaxation towards equilibrium will be governed by the rate constants for all the elementary individual steps which interconvert these species. If the original perturbation is made sufficiently small compared to the total concentrations of reactants, it is possible to obtain an exact solution to the kinetic equations which describe the relaxation process. Thus, after rapid perturbation of a simple two-state equilibrium:

$$\text{A} \underset{k_2}{\overset{k_1}{\rightleftharpoons}} \text{B}$$

the deviation ΔA in the concentration of A from its equilibrium value at time t will be given by

$$\Delta\text{A} = \Delta\,\text{A}_0 e^{-t/\tau} \qquad\qquad 15$$

where ΔA_0 is the original deviation at $t=0$. The parameter τ, the characteristic *relaxation time* for the system, may be evaluated directly from the observed time course of the relaxation. For this simple unimolecular process, it is related to the individual rate constants by

$$\tau = \frac{1}{k_1 + k_2} \qquad\qquad 16$$

For a system linked by n consecutive steps, there will be $(n-1)$ discrete relaxation times, each of which is a function of all the relevant rate constants, and, unless only unimolecular processes are involved, of the reactant concentrations too. It is correspondingly harder to interpret the chemical relaxation data for such systems. The number of relaxations observed often depends on the conditions of measurement, and if relaxation times are close together, they must be resolved by resort to a computer program. Results from relaxation studies are best interpreted in the light of what is known of the enzyme mechanism from steady-state and stopped-flow measurements.   Nevertheless, relaxation methods make a distinctive contribution, partly because of the detail of the information provided, but primarily because of the outstanding time resolution obtainable. In one widely-used technique, a temperature 'jump' of $10°K$ can be produced in a few micro-seconds by means of an electric discharge. The rate constant for the combination of an enzyme with a specific substrate can be accurately measured in this way, even if the combination rate approaches the limit set by diffusion in aqueous solution.

The effects of enzyme inhibitors

It is a convenient simplification to divide enzyme inhibitors into two classes, reversible and irreversible.   The usual criterion for reversibility is that the enzyme regains its original activity after dialysis to remove the inhibitor.   Clearly, though, if dialysis is not exhaustive, a tenaciously-bound reversible inhibitor may well seem to be irreversible in its action.

Reversible inhibition

In the simplest form of reversible inhibition, the inhibitor I has an affinity only for the free enzyme E, and the binding of the inhibitor and of the normal substrate is mutually exclusive.   This is termed *competitive inhibition*, because the substrate and the inhibitor compete for the free enzyme.   Competitive inhibitors which bear a structural resemblance to the substrate are usually assumed to compete directly for the active site.   On the other hand, the conformational flexibility of proteins makes it possible for an inhibitor which binds at a remote part of the enzyme surface thereby to distort the active site so that it is no longer accessible to substrate.

For enzyme reactions which follow Michaelis–Menten kinetics, a steady-state treatment shows that the addition of a competitive inhibitor has no effect on the maximal velocity V.   In molecular terms this is reasonable, because a substrate at high concentration will compete very effectively for the available enzyme.

At lower substrate concentrations, a proportion of the enzyme will be occupied in binding the inhibitor,

$$E + I \rightleftharpoons EI$$

the strength of this binding being expressed by the inhibition constant K_i:

$$K_i = \frac{[e][i]}{[ei]} \qquad 17$$

where [i] is the concentration of free inhibitor and [ei] is the concentration of inhibitor-enzyme complex.   The effect of this binding is revealed as an apparent increase in the value of K_m for the substrate.   If K_i can be evaluated for different competitive inhibitors, useful information is provided about the structural requirements for binding at the active site.

A *non-competitive* inhibitor is defined as one which can bind both to the free enzyme and to the enzyme-substrate complex, in such a way that substrate and inhibitor binding are mutually independent. The kinetic equations for the apparent Michaelis parameters become splendidly complex, but two main points of importance emerge; first, V is always reduced and the inhibition cannot be overcome by high concentrations of substrate; secondly, for systems where the breakdown of the ES complex into products is clearly the slowest step, the K_m of the substrate is unchanged. There are a large number of enzyme-catalysed reactions where a given inhibitor does not fall obligingly into one or other of these categories, and in such a case there are apparent changes in both V and K_m.

A clear example of competitive inhibition is the action of malonate on the oxidation of succinate to fumarate by succinate dehydrogenase (*Fig. 26*). Evidently the enzyme's active site is unable to discriminate effectively against malonate because of its close structural resemblance to the true substrate.

Irreversible inhibition

Irreversible inhibitors, almost without exception, act by covalent modification of the enzyme protein. Denaturing agents such as urea also abolish enzyme activity, but this is because they unfold the protein and such effects are non-specific and usually require large molar excesses of reagent.

The amino acid residues in proteins which are most susceptible to covalent modification are those with neutral polar or charged polar side-chains, such as tyrosine, aspartic acid, cysteine and lysine. It would obviously require an inhibitor of exceptional chemical reactivity to modify a residue having a hydrocarbon side-chain such as alanine or leucine. Other generalisations are of limited use,

FIG. 26. Malonate competitively inhibits succinate dehydrogenase because of its close structural resemblance to the substrate.

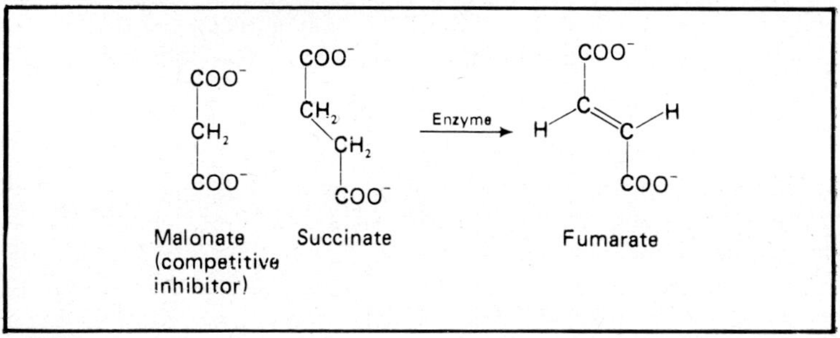

Malonate (competitive inhibitor) Succinate Fumarate

because the chemical reactivity of each side-chain is so profoundly influenced by its local environment in the enzyme protein. In other words, if a reagent which reacts specifically with histidine residues in small peptides is found to inactivate an enzyme, this is only presumptive evidence for the involvement of histidine in that enzyme's catalytic action. The histidine residue may not be at the active site, and indeed the susceptible side-chain may well turn out not to be histidine at all since some other residue may possess enhanced susceptibility towards the reagent used. Another problem is that even if only an equimolar amount of inhibitor is used, various amino acid side-chains may compete successfully for a share, and a mixture of protein species may be obtained, all with different levels of enzyme activity. Fortunately, most such pitfalls in the interpretation of covalent enzyme inhibition may be avoided by linking the kinetic studies to a thorough structural characterisation of the modified protein.

Covalent inhibitors have been useful in identifying individual residues which are important for catalysis. Two basically different approaches can be used: one relies on amino acid side-chains in the enzyme active site being uniquely reactive towards an inherently unselective reagent; the other, in contrast, employs reagents that closely resemble the natural substrate and are therefore specifically bound at the active site. The latter technique is often referred to as affinity labelling. Examples of both approaches are given in Chapter 4.

Results from chemical modification experiments, particularly with unselective reagents, must always be treated with caution. Iodoacetate, for example, is often used to block cysteine residues, but may also react with methionine, lysine and histidine. Alternatively, no modification may be found, although the protein is known to contain susceptible groups. It used to be assumed that such residues must be 'buried' in the protein interior, but x-ray structural work has shown that even apparently 'exposed' residues may be inert. Much remains to be learned about the chemical reactivity of protein side-chains. In general, negative results are the most decisive. If an enzyme residue is modified without catalytic activity being affected, then that particular residue is clearly not directly involved in the catalysis.

Inhibition involving metal ions

A large number of enzymes are rapidly inhibited by traces of certain metal ions, notably copper, mercury, lead and silver. Nucleophilic amino acid side-chains readily act as ligands towards these metals forming a covalent complex, and if a critical active site residue is involved, then the site is blocked and the enzyme poisoned. Mercury

Hg(II), lead Pb(II) and cadmium Cd(II) ions form very stable complexes with sulphur-containing ligands, and they are often found to inhibit enzymes which contain free thiol groups.   In one or two cases the nature of the complex has been established by x-ray crystallographic analysis at high (0.2 nm or 2 Å) resolution; for example, Hg(II) ions bind specifically to the active site of bacterial glyceraldehyde 3-phosphate dehydrogenase.   The heavy metal ion binds behind the active site, and the catalytically essential residue cysteine-149, together with a second residue cysteine-151, are liganded to the metal through their sulphur atoms.   In this example, the stability of the enzyme-inhibitor complex is increased by a chelation effect, since more than one amino acid residue is involved.   The entropic advantage here is exactly the same as that which underlies the stability of metal ion complexes with EDTA (ethylenediaminetetraacetic acid).

The susceptibility of thiol-containing enzymes and co-enzymes to inhibition by heavy metal ions partly accounts for the cumulative toxicity of these elements.   Another point of practical importance is that even highly purified water and laboratory chemicals may contain enough heavy metal ions to inactivate sensitive enzymes which (because of their high relative molecular mass) are usually present in low molar concentrations.   To prevent this, it is often necessary to sequester the metal ions by addition of a suitable chelating agent. Conversely, chelating agents, and other ligands such as cyanide or fluoride, inactivate a number of enzymes.   This is an indication (though by no means a proof) that such enzymes contain one or more firmly-bound metal atoms which are essential for either the structural integrity of the protein, or for the catalysis itself.

4. Towards an Enzyme Mechanism

The aim of this Chapter is to demonstrate, using two examples, how information from different experimental approaches can be integrated into a detailed description of an enzyme-catalysed reaction. Such a description, as we shall see, provides an exciting insight into the chemical basis of enzyme catalysis.

Chymotrypsin

Chymotrypsin is a small extracellular enzyme, with a relative molecular mass of about 25 000, which hydrolyses food proteins in the mammalian gut. It is synthesised and stored as chymotrypsinogen, an inactive precursor containing 245 amino acids in a single polypeptide chain. Trypsin, a closely-related protease, 'switches on' chymotrypsin by specific hydrolysis of a single internal peptide bond in the precursor, and several peptide bonds are then attacked by chymotrypsin itself. In the commonest form of the enzyme, α-chymotrypsin, there are three separate chains resulting from the excision of two dipeptides from the precursor, and representing residues 1–13, 16–146 and 149–245 of the original chymotrypsinogen molecule. The three chains are held together by –S–S– bonds, and by non-covalent interactions. The complete three-dimensional structure of α-chymotrypsin has been worked out using x-ray crystallography, and the folding of the polypeptide backbone is shown in *Fig. 27*, with the amino acid side-chains omitted.

The molecule is roughly spherical and extremely compact, with the main polypeptide chains folded closely back upon one another. Folding is rather irregular, although there are some short lengths of β-sheet secondary structure, and a small amount of α-helix, of which the segment from residues 234–245 is the most visible. The enzyme requires no metal ions or cofactors for its activity.

Specificity of action

Chymotrypsin catalyses the hydrolysis of internal peptide links in natural polypeptide chains, a thermodynamically favourable process that goes essentially to completion:

$$\text{\Large \wedge\wedge-NH-CH-C-NH-CH-C-}\wedge\wedge + H_2O \rightleftharpoons$$

$$\wedge\wedge-NH-CH-C-OH + NH_2-CH-C-\wedge\wedge$$

46

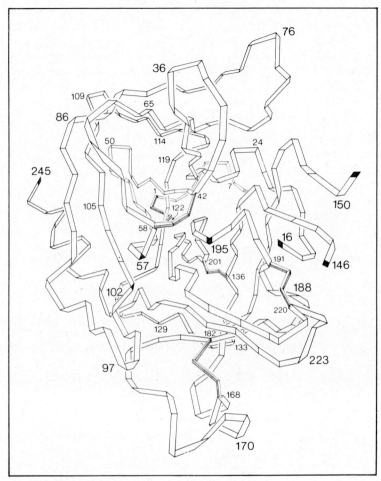

FIG. 27. The folding of the polypeptide backbone of α-chymotrypsin. Only the positions of the α-carbon atoms are shown. (From D. Blow in *The enzymes*, vol. 3, 3rd edn. London: Academic, 1971.)

Chymotrypsin also catalyses the hydrolysis of 'model' substrates such as *N*-acetyl-L-phenylalanine ethyl ester:

$$CH_3{-}\overset{O}{\overset{\|}{C}}{-}NH{-}\underset{L}{\overset{Phe}{\overset{|}{C}H}}{-}\overset{O}{\overset{\|}{C}}{-}O{-}CH_2CH_3 + H_2O \rightleftharpoons$$

$$CH_3{-}\overset{O}{\overset{\|}{C}}{-}NH{-}\overset{Phe}{\overset{|}{C}H}{-}\overset{O}{\overset{\|}{C}}{-}OH + CH_3CH_2OH$$

and of many other synthetic esters, amides and peptides. These synthetic substrates have been of great value in delineating the structural requirements for fast hydrolysis. Chymotrypsin, it turns out, preferentially attacks bonds on the carboxyl side of bulky, non-polar amino acid residues, such as phenylalanine, tyrosine or tryptophan. Further, derivatives of L-amino acids are hydrolysed much faster than their unnatural counterparts with the D-configuration.

The acyl-enzyme intermediate

With a wide range of substrates, chymotrypsin-catalysed hydrolysis obeys the Michaelis–Menten kinetic equation, so that it is safe to assume the initial formation of an enzyme-substrate complex. Subsequent breakdown into products appears to take place in at least two steps. The first evidence for this came from rapid reaction studies using the synthetic substrate p-nitrophenyl acetate. When p-nitrophenyl acetate is mixed with chymotrypsin, the release of p-nitrophenol is seen to occur in two phases, a rapid 'burst' followed by relatively slow, steady reaction (*Fig. 28*).

The amplitude of the initial burst corresponds to the liberation of one mole of phenol for each mole of enzyme. The simplest interpretation is that hydrolysis involves two separate chemical steps. In the first step, which is rapid, the acetyl group is transferred to the enzyme, liberating p-nitrophenol (*Fig. 29a*). In the second, slower step the acetyl group is transferred to a water molecule, regenerating the free enzyme (*Fig. 29b*).

In the 'burst' phase of the hydrolysis, as the acetyl-enzyme intermediate is accumulating, the rate of phenol production is determined by the acetylation step. Thereafter, in the steady state, phenol release and the rate of the overall reaction are both limited by the slow de-acetylation of this intermediate. Further work has established that all other ester substrates of chymotrypsin are hydrolysed by way of an acetyl-enzyme (or, in general, an acyl-enzyme) intermediate. The experimental rate constant k_{cat} obtained from steady-state kinetic measurements is, to a first approximation, identical with the rate constant for de-acetylation, so that esters which generate the same acyl–enzyme have the same overall rate constant, k_{cat}, for hydrolysis.

An acyl-enzyme is also involved in the hydrolysis of amide and peptide substrates, but with one major difference: in this case the *formation* of the intermediate is the slowest step, so that no analogous 'burst' of amine production can be detected. This change in rate-determining step is consistent with the known chemistry of amides and esters, in that amides are hydrolysed (by OH^-, at least) about 1000 times slower than the corresponding esters.

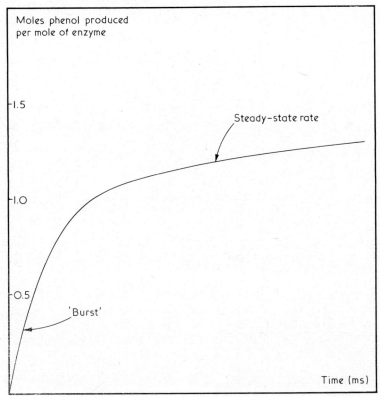

FIG. 28. The reaction of *p*-nitrophenyl acetate with α-chymotrypsin.

Identification of enzyme groups directly involved in catalysis

Since the acyl-enzyme intermediate in amide and peptide hydrolysis is formed after the rate-limiting step, it can only be detected by indirect kinetic methods. On the other hand, the same acyl-enzyme generated during the hydrolysis of an ester does reach an appreciable concentration during the steady state phase and, if the reaction mixture is rapidly acidified to pH 4, the intermediate becomes stable enough to be studied directly. To identify the point of attachment of the acyl group to the enzyme, carbon-14 labelled *p*-nitrophenyl acetate was used to generate a radioactive acyl-enzyme. The modified protein was then carefully degraded, and all the radioactivity was found associated with a unique amino acid residue, serine-195 (the numbering refers to the single chain of chymotrypsinogen). The initial acyl-transfer may, therefore, be written as in *Fig. 30a*. This result is rather surprising at first sight, because the hydroxyl group in the side-chain of serine is not a very reactive nucleophile,

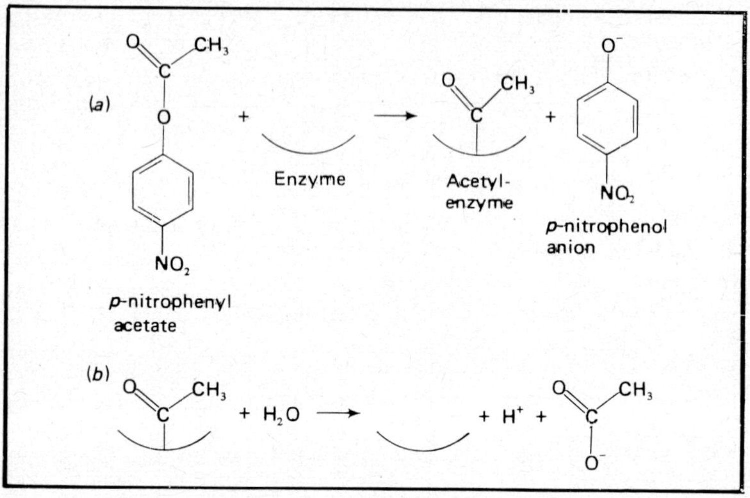

FIG. 29. The proposed two-step mechanism for hydrolysis of *p*-nitrophenyl acetate, by way of an acetyl-enzyme intermediate.

and because there are 27 other serine residues in the molecule that remain untouched. The special reactivity of serine-195 is confirmed by the finding that organophosphorus compounds such as diiso-propylphosphofluoridate react quantitatively with chymotrypsin and completely destroy the enzyme activity (*Fig. 30b*). Here, too, serine-

FIG. 30. The reaction of serine-195 in α-chymotrypsin with (*a*) the substrate *p*-nitrophenyl acetate and (*b*) the irreversible inhibitor diisopropylphospho-fluoridate.

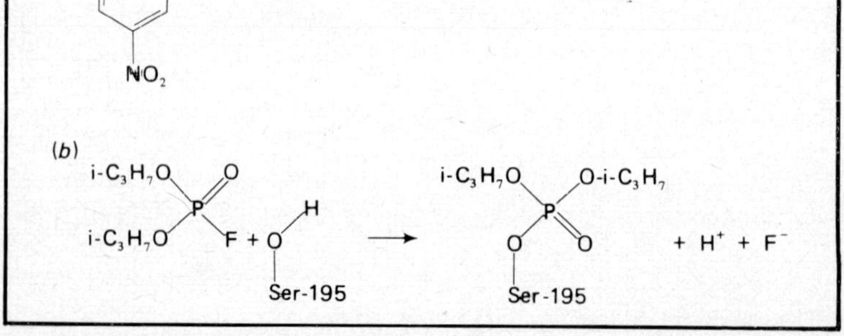

195 is the unique site of attachment, and since the phosphoryl-serine linkage is stable, the inhibition is essentially irreversible. Due to an analogous effect on the enzyme acetylcholinesterase in nerve endings, many such organophosphorus compounds are potent nerve gases.

Once the special role of Ser-195 had been established, attempts were made to identify other enzyme groups at the active site. One such experiment is a classical example of the approach to specific chemical modification called 'affinity labelling'. Compare the structures of the two derivatives of L-phenylalanine in *Fig. 31*. The first (*a*) is an excellent substrate for chymotrypsin, the second (*b*) is not, but does contain many of the same structural features and should possess a certain 'affinity' for the enzyme active site. Once delivered to the active site, any suitably-positioned nucleophilic amino acid could easily displace chloride ion from this reactive α-chloroketone. In practice, this substrate analogue was indeed found to be an efficient, irreversible inhibitor of chymotrypsin. It was subsequently shown that the inhibition was due to the specific modification of a basic active site residue, histidine-57, acting in this case as a nucleophile. This finding seemed to confirm the suggestion that His-57 was responsible for the special nucleophilic reactivity of Ser-195, through its ability to act as a general base, and deprotonate the serine hydroxyl group (*Fig. 32*).

The substrate binding site

In the three-dimensional structure of the enzyme, the substrate binding site is close to the uniquely reactive serine-195. A deep, but

FIG. 31. Comparison of the structure of a good substrate (*a*) for chymotrypsin with that of an active site-directed irreversible inhibitor (*b*).

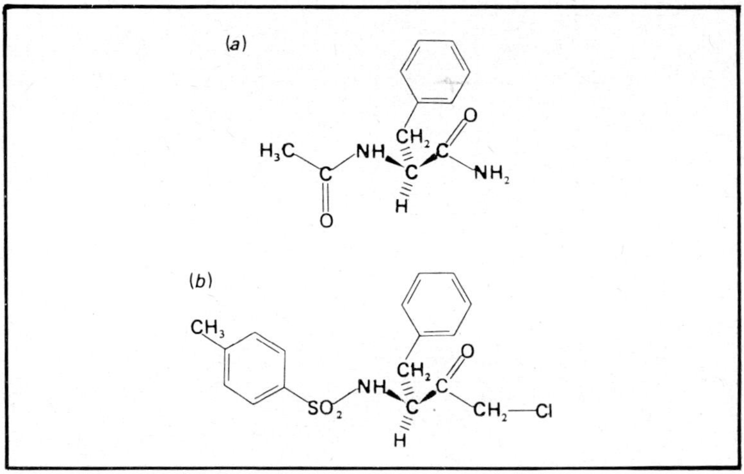

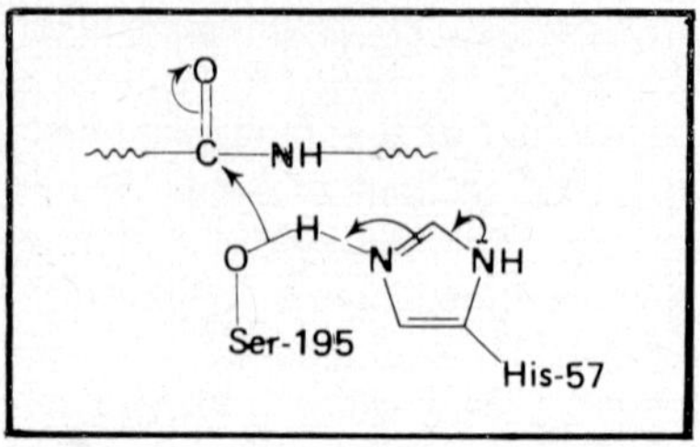

Fig. 32. Proposed general base catalysis by histidine in chymotrypsin hydrolysis.

rather narrow, pocket created by the folding of the enzyme chain, lined with largely hydrophobic residues, is designed to accommodate the non-polar side-chain of a typical substrate. This favourable interaction provides much of the energetic driving force for substrate binding, but it also serves to position the substrate at the active site. Specificity studies indicate that a second favourable interaction involves the acylamino portion of the substrate, and from the crystal structure this can be identified as a hydrogen bond to the peptide carbonyl group of Ser-214. The susceptible carbonyl group points towards the main-chain peptide groups of both Gly-193 and Ser-195, which play an important role in the catalysis. These interactions are summarised, for a peptide substrate containing L-phenylalanine, in *Fig. 33*.

If these precise interactions are not made, the susceptible carbonyl group will be oriented wrongly, away from the enzyme nucleophile Ser-195. This explains the high stereospecificity of the enzyme, for both D- and L-amino acid derivatives can be tightly bound, but only L-amino acid derivatives can adopt the conformation required for reaction.

The charge-relay system

The active site of chymotrypsin can be located in the three-dimensional structure as a shallow depression on the enzyme surface, next

Fig. 33. Some important binding interactions at the active site of α-chymotrypsin. The point of attack by Ser-195 is arrowed.

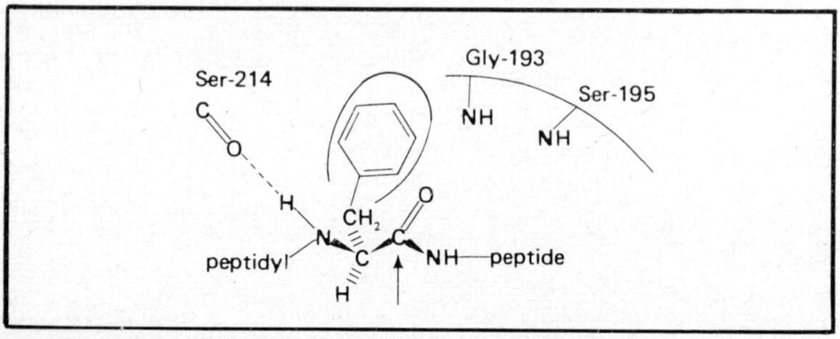

to the substrate binding site. As the chemical studies predicted, it contains both serine-195 and histidine-57, and the hydroxyl group of the serine is sufficiently close to the imidazole ring of the histidine that it forms a hydrogen bond to one of the ring nitrogen atoms. The second ring nitrogen is in turn hydrogen-bonded to the negatively-charged carboxylate group of aspartic acid-102, which lies directly behind it, completely screened from the solvent and in a rather hydrophobic environment. This would not have been easy to predict, because fully-charged polar side-chains are expected to be rather uncomfortable in such a non-polar environment, but aspartate-102 is thereby ideally suited to act as a general base. Protonation of the buried negative charge should, of course, be a particularly favourable process. The importance for the catalysis of this so-called charge-relay system is underlined by the discovery that the bacterial protease subtilisin contains the same three residues, a serine, a histidine and an aspartic acid, at its active site, although it shows no other structural resemblance to chymotrypsin. It is as though nature has found the same solution, to the problem of how to hydrolyse a peptide bond, by two independent routes. A fuller description of the critical acylation step, showing the operation of the charge-relay system, is given in *Fig. 34*.

The nucleophilic attack by Ser-195 on a peptide substrate leads first to an unstable adduct with tetrahedral geometry. As the tetrahedral intermediate collapses to the acyl-enzyme, proton transfer from the charge-relay system allows the amine to depart as a neutral species, and restores His-57 and Asp-102 to their original ionisation state. In the next step, a water molecule (shown hydrogen-bonded to the acyl-enzyme) will attack the carbonyl group and bring about de-acylation.

The rate of the initial attack by serine-195 determines the overall rate of hydrolysis for peptide substrates. It is, therefore, interesting to see that this step is assisted in at least three ways: by the use of precise binding interactions to orient the reaction components in the optimal position for reaction: by the operation of the charge-relay system (which might more appropriately be called a proton relay system); and by the stabilisation of the transition state, in which negative charge develops on the carbonyl oxygen atom, through perfectly placed hydrogen bonds from the main-chain peptide nitrogen atoms of glycine-193 and serine-195.

Alcohol dehydrogenase

The reversible oxidation of ethanol into acetaldehyde is catalysed in liver tissue by the enzyme liver alcohol dehydrogenase (LADH).

$$CH_3CH_2OH \rightleftharpoons CH_3CHO$$

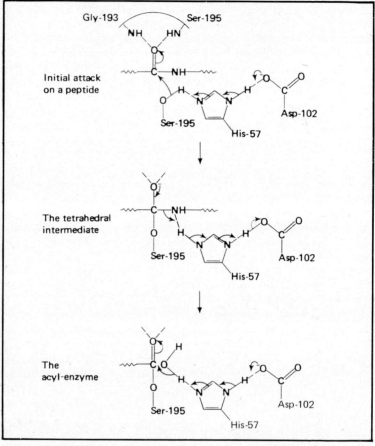

FIG. 34.   The acylation step in chymotrypsin-catalysed peptide hydrolysis.   The departing amine is replaced by a water molecule which brings about the subsequent deacylation.

The enzyme requires a co-enzyme for activity, nicotinamide adenine dinucleotide (NAD$^+$), which acts as a biological oxidising agent.

The role of the co-enzyme

The co-enzyme is shown in *Fig. 35* in its oxidised form (NAD$^+$). The oxidation–reduction properties are solely due to the nicotinamide portion of the molecule.   During the LADH-catalysed reaction, a hydrogen atom is transferred directly, with both its bonding electrons, from the alcohol to the C-4 position of the nicotinamide ring, as shown in *Fig. 36*.   Acetaldehyde (ethanal), a proton and the reduced form of the co-enzyme (NADH) are produced in equimolar

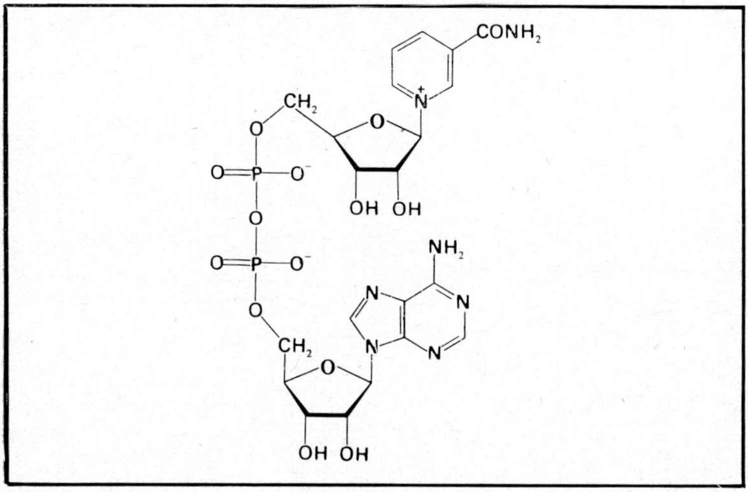

FIG. 35. The structure of nicotinamide adenine dinucleotide in its oxidised form (NAD$^+$).

amounts. This reaction is infinitely slow in the absence of LADH, so clearly the enzyme active site must create a peculiarly favourable environment for this hydride transfer.

The control exerted by the enzyme is shown most clearly by the stereospecificity of the hydrogen transfer, as revealed by isotope labelling experiments. As indicated in *Fig. 36*, the hydrogen abstracted from ethanol (H$_A$) is one of a pair of chemically identical atoms. These atoms, however, are sterically non-equivalent (they are mutually related in the same way as the hydroxymethene groups of glycerol) and in the ordered, chiral environment of the enzyme's active site they are readily differentiated. The stereospecificity of the transfer also extends to the cofactor: H$_A$ always migrates specifically to one face of the nicotinamide ring, as indicated in *Fig. 36*. The paired hydrogen atoms at the C-4 position in NADH

FIG. 36. The stereochemistry of the hydrogen transfer in the oxidation of ethanol by LADH.

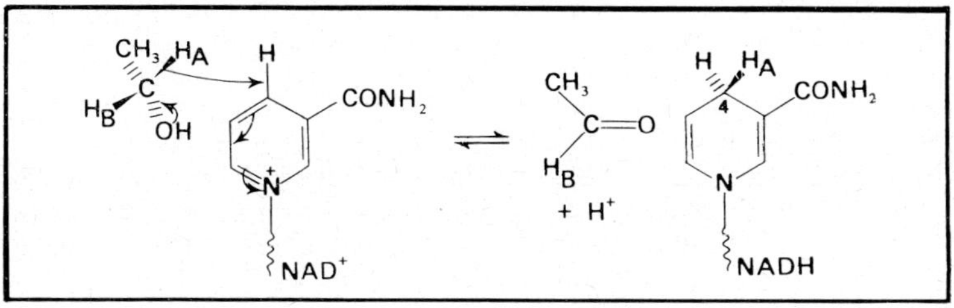

are also geometrically non-equivalent (they can be distinguished by selective achiral reagents, as well as by an enzyme) so that when this NADH is used in the reverse reaction, only H_A is transferred to acetaldehyde.

Kinetic evidence

On the basis of their steady state kinetic studies, Theorell and Chance have proposed the following pathway for the oxidation of ethanol and other primary alcohols:

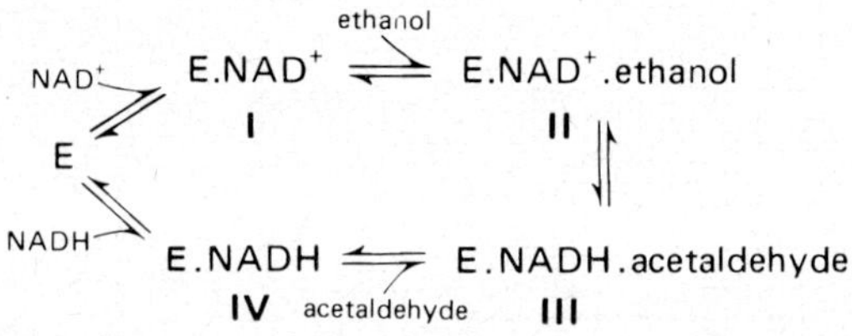

One essential feature of this pathway is that hydrogen transfer occurs within a ternary complex containing enzyme, substrate and co-enzyme. Another is the existence of a compulsory order of binding: NAD+ always binds before ethanol, NADH always binds before acetaldehyde. This simple scheme accounts rather well for the kinetic data, when allowance is made for so-called 'dead-end' ternary complexes, formed when ethanol binds to binary complex IV, or when acetaldehyde binds to binary complex I. The compulsory binding order clearly suggests that co-enzyme binding enforces some change in the enzyme structure that enables the substrate to bind, as visualised by Koshland in his 'induced fit' mechanism.

Another point of general interest has emerged from kinetic studies using fast reaction techniques: in the oxidation of substrates such as ethanol, the overall rate is governed by the rate of desorption of NADH from the enzyme. Hydrogen transfer within the ternary complex is evidently so efficient that the actual catalytic step is faster than the step involving NADH release.

Substrate specificity

LADH has a rather broad substrate specificity and, apart from ethanol, many primary and some secondary alcohols are oxidised, to aldehydes and ketones respectively. The reactions shown in *Fig. 37* are representative. A comparison of the rate constants for the oxidation of different substrates shows clearly that the active site can accommodate quite bulky alcohols, although there is a distinct preference for substrates containing hydrophobic groups.

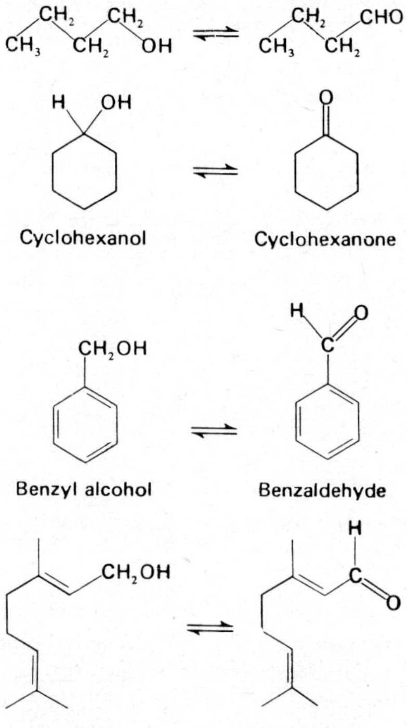

FIG. 37. Representative alcohol substrates for liver alcohol dehydrogenase.

Active site groups involved in catalysis

The LADH molecule is a dimer, with a relative molecular mass of 80 000. The subunits are identical and each contains 374 amino acid residues in a single polypeptide chain, as well as two firmly bound zinc atoms. One of the zinc atoms in each subunit is clearly at the active site. When it is removed from the protein, enzyme activity is lost, and when zinc-complexing agents such as 1,10-phenanthroline are present, they act as competitive inhibitors towards substrate and bind near this zinc atom. The second metal atom is not concerned directly with catalysis, and its precise function is not known.

The importance of the 'catalytic' zinc atom at the active site is underlined by the results of chemical modification studies. Of a total of 14 cysteine residues in each subunit, one is unusually reactive towards the non-selective reagent iodo-acetic acid. The reactive residue is cysteine-46, one of three protein side-chains directly liganded to the active site zinc. A second ligand, cysteine-174, has been identified as the site of selective modification by a reactive co-enzyme analogue. The involvement of further protein residues in the special chemistry of the active site was only clarified by the determination of the full three-dimensional structure by x-ray crystallography.

The structure of LADH

The structure of the horse liver enzyme has been determined to high resolution by Brändén's group at Uppsala, using x-ray crystallography. Actually, independent determinations were undertaken not only of the free enzyme, but also of various binary complexes with co-enzyme, co-enzyme analogues and competitive substrate inhibitors. The entire amino acid sequence has also been determined, and this has helped to reveal the positions of nearly all the amino acid side-chains.

The schematic cross-section in *Fig. 38* shows the gross features of the enzyme structure. Each subunit is itself composed of two clearly separated domains, the smaller of which contains both the co-enzyme binding site and the interface region where the subunits are held together by non-covalent interactions. The outer domain contains both the zinc atoms, one in the neck region which connects the two parts of the subunit, and the other (the 'catalytic' zinc) on the inner face, near the junction with the other subunit. In this position

FIG. 38. A schematic cross-section of the liver ADH molecule, viewed along the rotational axis of symmetry. (From C. I. Brändén in *Pyridine nucleotide dependent dehydrogenases*, H. Sund (ed). Heidelberg: Springer Verlag, 1977.)

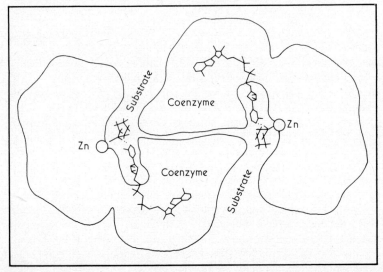

the catalytic zinc atom is only accessible from two directions. A narrow crevice above allows the co-enzyme, when fully extended, to point down into the active site region. The other direction of approach is at right angles to the first, a deep pocket where, in *Fig. 38*, the substrate cyclohexanol is shown binding directly to the zinc. *Figure 39* presents a stereo diagram of one subunit of LADH from the opposite side of the molecule to that represented in the schematic diagram of *Fig. 38*. Only the α-carbon atom of each residue is shown, together with the position of the zinc atoms. Compared to α-chymotrypsin, the folding of the polypeptide chain shows a great deal of intricate secondary structure, particularly in the co-enzyme binding domain.

The active site

A schematic diagram of the active site of LADH, as deduced from the x-ray studies, is shown in *Fig. 40*. The catalytic zinc atom lies at the bottom of the deep substrate-binding pocket, almost 2.5 nm (25 Å) from the enzyme surface. The side-chains Cys-46, Cys-174 and His-67 are all liganded to the metal, and an approximately tetrahedral coordination is completed by a water molecule. The oxygen atom of the zinc-bound water molecule forms a hydrogen bond to the hydroxyl group of Ser-48, which is in turn hydrogen-bonded to the side-chain of His-51 in a manner reminiscent of the 'charge relay' system in chymotrypsin. This hydrogen-bond network is shown more clearly in the stereo diagram of *Fig. 41*.

The adenine portion of the co-enzyme binds snugly into a hydrophobic pocket on the smaller domain of each subunit, and the rest of the molecule extends into the crevice which leads to the active site, so that the crucial C-4 position of the nicotinamide ring is brought to within 0.45 nm (4.5 Å) of the catalytic zinc, with one face specifically turned towards the metal. In a conformation where the other face of the nicotinamide group is turned towards the zinc, there is steric interference between the amide group of the nicotinamide and the side-chain of Val-203.

A hydrogen bond from the hydroxyl group of Thr-178 to the carboxamide side-chain of the co-enzyme, ensures accurate positioning of the nicotinamide ring in the active site cleft.

Figure 40 also shows that the substrate, in order to reach the active site, must pass through a tube or barrel which is formed by the uniformly hydrophobic side-chains of the residues which line the substrate binding pocket. This hydrophobic region is presumably the key to the observed preference for non-polar substrates. A substrate's eye view, down through this barrel towards the zinc, is provided in the stereo diagram of *Fig. 42*. Residues Ser-48, Leu-57, Val-58, Phe-93, Phe-110, Leu-116, Phe-140, Leu-141, Pro-296 and

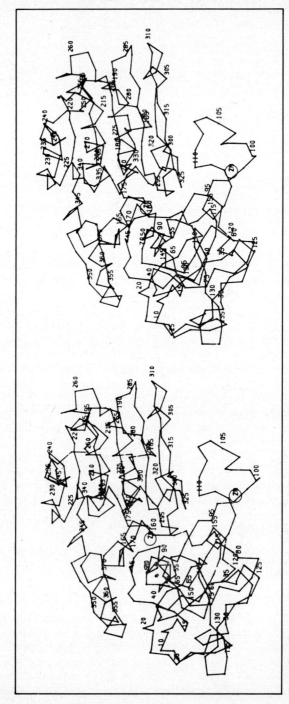

FIG. 39. Computer-drawn stereo diagram of the polypeptide chain of one subunit of LADH. Only the α-carbon atoms and the zinc atoms are shown. (From C. I. Brändén in *The enzymes*, vol. 3, 3rd edn. London: Academic, 1971.)

40. Schematic diagram of the active site of LADH as deduced by x-ray
allography. (From C. I. Brändén in *Pyridine nucleotide dependent dehydro-*
ses H. Sund (ed). Heidelberg: Springer Verlag, 1977.)

. 41. Stereo diagram showing the spatial orientation of the zinc-bound water
lecule, the side-chains of Ser-48 and His-51 and the hydrogen bond system
tween these groups. (From H. Eklund, *J. mol. Biol.*, 1976, **102**, 27.)

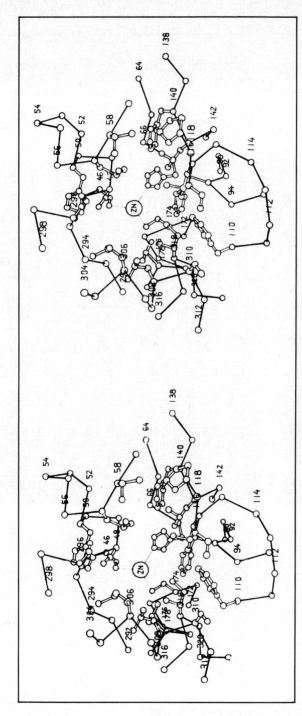

Fig. 42. Stereo diagram showing the view down through the 'hydrophobic barrel' towards the active site zinc. (From C. I. Brändén in *The enzymes*, vol. 3, 3rd edn. London: Academic, 1971.)

Ile-318 are contributed from the outer domain of one subunit, Met-306 and Leu-309 from the co-enzyme binding domain of the other subunit.

A possible mechanism of action

There is spectroscopic evidence that the oxygen atom of substrate alcohols binds directly to the catalytic zinc atom in LADH. In agreement with this, the specificity of the enzyme towards a variety of cyclohexanol derivatives can be neatly correlated with the geometry of the substrate binding site deduced from the x-ray work, if direct binding to the metal is assumed, and the hydrogen to be transferred points towards the C-4 atom of the co-enzyme nicotinamide ring. It has, therefore, been proposed that the role of zinc is to polarise the alcohol (an oxygen–zinc alkoxide bond has electron-donating properties). The catalysis becomes a simultaneous electron 'pull' from the positively charged nitrogen atom of nicotinamide and a 'push' from the alkoxide bond, as in *Fig. 43a*. In the reverse reaction, the free electron pair on the nicotinamide ring nitrogen provides the 'push' and the zinc atom, by coordinating the carbonyl group and enhancing its acceptor properties, provides the 'pull'.

Is such a mechanism reasonable in chemical terms? A pertinent chemical reaction is the Oppenauer oxidation of primary and secondary alcohols by acetone, in the presence of a catalytic amount of aluminium isopropoxide. As indicated in *Fig. 43b*, the reaction probably involves direct hydride transfer within a complex where the aluminium serves both to activate the alcohol and to polarise the

FIG. 43. (*a*) A possible mechanism for hydrogen transfer at the LADH active site and (*b*) a chemical model reaction, the Oppenauer oxidation.

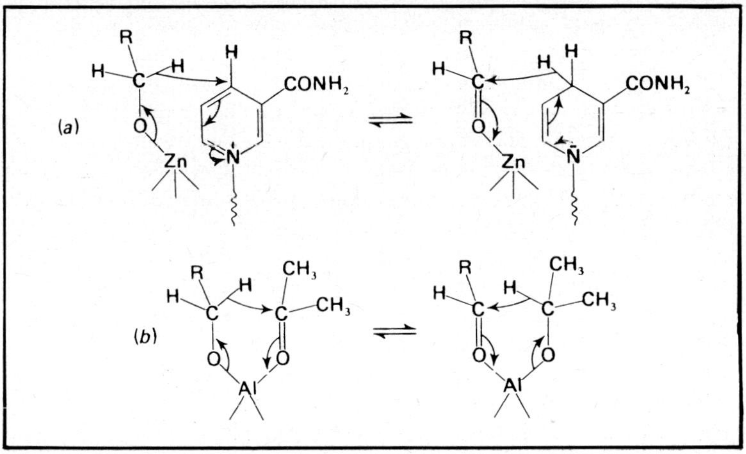

carbonyl group of the ketone. The analogy with the LADH-catalysed reaction seems attractive.

This is not, however, quite the whole story. One objection is that although in the Oppenauer oxidation the intermediate alkoxide is easily formed by an exchange reaction:

$$R—CH_2OH + Al-[-OCH(CH_3)_2]_3 \rightleftharpoons$$
$$R—CH_2O—Al-[-OCH(CH_3)_2]_2 + (CH_3)_2CHOH$$

considerable energy would be required to deprotonate a neutral alcohol to form an alkoxide at the LADH active site (the pK_a of ethanol is about 18). Secondly, when the alcohol structure is varied systematically, by introducing electron-withdrawing or electron-donating groups, the relative rates of reaction indicate that no net charge develops on the substrate during the oxidation. This implies that hydride transfer to the co-enzyme, and loss of the proton from the hydroxyl group, may occur simultaneously.

A possible way of resolving these difficulties is suggested by the results of pre-steady state kinetic studies. When the oxidised form of the co-enzyme (NAD$^+$) binds to the enzyme, protons are immediately released into the solution, because the pK_a of an active site group is perturbed from about 9.0 to about 7.6. This active site group has been identified as the water molecule bound to the catalytic zinc, which because of its coordination to the metal is much more acidic than a free water molecule (pK_a 15.7). At neutral pH values, the binary enzyme-NAD$^+$ complex will, therefore, largely contain a hydroxide ion attached to the zinc. The substrate, it has been found, binds specifically to this form so that (in the presence of saturating amounts of alcohol) the deprotonation of the zinc-bound water molecule is driven to completion by the formation of the reactive ternary complex. Since this proton release on binding occurs during every 'turnover' of the enzyme, a proton appears in the overall stoichiometry of the oxidation. It is possible that the deprotonation is speeded up by exploiting the hydrogen-bonding network which connects the zinc-bound molecule to the hydroxyl group of Ser-48 and the imidazole group of His-51. This constitutes a *proton relay* which could ensure, as shown schematically in *Fig. 44*, that protons are released at the polar enzyme surface rather than into the hydrophobic active site.

This kinetic information makes it very probable that it is the hydroxide ion bound to the catalytic zinc which provides electron 'push' for the hydride transfer step by acting as a general base (*see Fig. 45*). In the reverse reaction, in accord with the principle of microscopic reversibility, a zinc-bound water molecule acts as a general acid. In this modified mechanism the zinc atom has two

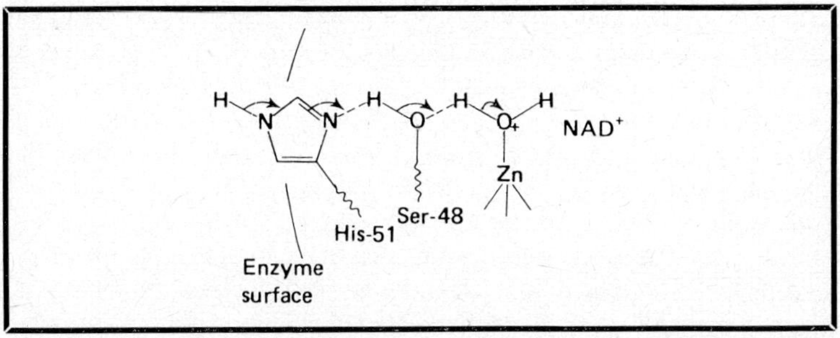

FIG. 44. A possible 'proton relay' system in LADH, channelling protons from the relatively hydrophobic active site to the enzyme surface.

separate functions: to orient and polarise the substrate and to enhance the reactivity of the bound water molecule.

Summary

The work on α-chymotrypsin and LADH illustrates the progress which has been made towards understanding enzyme catalysis. In each case we have a detailed kinetic scheme, with the structures of enzyme-bound intermediates and a knowledge of the rate constants of the individual steps which interconvert them. The high-resolution x-ray crystal structure suggests the involvement of certain enzyme functional groups and (equally important) enables many otherwise plausible mechanisms to be discarded. The essential

FIG. 45. The revised mechanism for hydrogen transfer at the LADH active site.

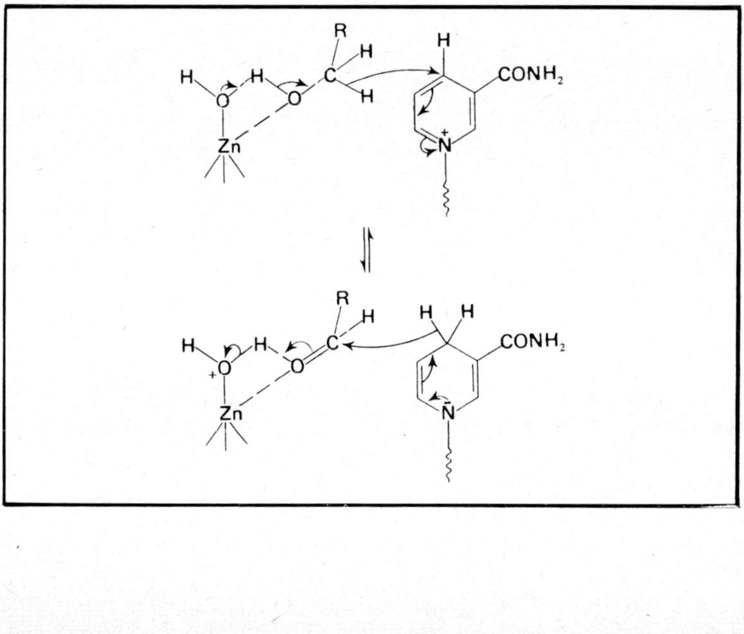

bridge between kinetic and structural studies is provided by chemical modification experiments which reveal details of the chemical reactivity of active site functional groups, and by stereochemical information which further restricts the number of possible mechanisms. Experiments on appropriate model systems help to integrate enzyme action into the framework of physical organic chemistry.

The description of catalysis which emerges is still incomplete, and is still being actively refined. Where, as with chymotrypsin and LADH, the information is detailed enough, it is exciting to find direct evidence for catalytic effects of the type discussed in Chapter 2. As more enzymes are studied, and as experimental techniques improve, the search for coherent general principles of enzyme action is showing itself to be increasingly rewarding.

5. Enzyme Technology and its Applications

The properties of enzymes as biological catalysts have long been relied upon in the traditional fermentation and brewing industries, but only recently has their potential been recognised as more broadly-based industrial catalysts. Where enzymes can be used rather than conventional chemical catalysts, the reaction often proceeds faster, more efficiently and under milder conditions.

At the same time, there are formidable technical problems to be solved in order to make enzyme-based processes feasible and commercially attractive, particularly for large-scale operations. These problems include the cost of purified enzymes, the possibility of microbiological contamination, and, for many enzymes, the cost of essential cofactors. For some enzymes of potential interest, the substrate specificity is so rigidly defined that they cannot be coaxed into accepting other, closely related, compounds. Finally, many purified enzymes are inherently unstable and are rapidly denatured under the reaction conditions.

These general limitations on enzymes as catalysts have been tackled in several ways. The most important of these involves the technique of enzyme 'immobilisation', but new methods for the more efficient production and stabilisation of enzymes are also being actively developed. Some of this work will be described here, as it provides a useful framework for the discussion of individual applications of enzymes.

Immobilisation methods

The term *immobilisation* implies the artificial confinement of the enzyme protein so that it remains accessible to its substrates, but is readily separable from them. In this way an enzyme may be re-used in a batch process, or even incorporated into a continuous reactor. Reaction can be rapidly initiated or quenched, and the products are obtained catalyst-free. In addition, immobilised enzymes are often more stable to denaturation than the free species. *Fig. 46* shows four typical ways in which immobilisation can be brought about.

Covalent bond formation

The most widely used method of immobilising enzymes involves a chemical reaction between amino acid side-chains on the protein and a suitably activated solid support. Hundreds of chemical coupling procedures have now been described, involving both synthetic solid supports such as polyacrylamide and nylon, and natural materials such as cellulose, agarose and other polysaccharides.

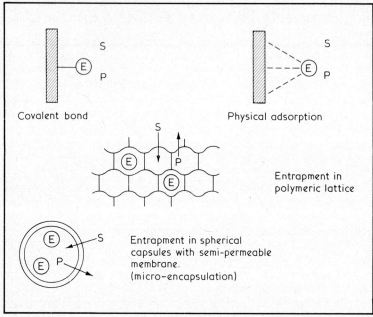

FIG. 46.   Four typical methods for immobilisation of an enzyme E, that catalyses the reaction S ⇌ P.

Immobilisation is successful if the coupling reaction is efficient, if the protein binding capacity of the water-insoluble support proves to be high, and if the immobilised enzyme retains activity.   Since the covalent attachment involves chemical modification of the protein, it is not surprising that, in general, the specific activity of the immobilised enzyme is somewhat less than that of the native enzyme in free solution.   For macromolecular substrates, this may partly reflect slow diffusion rates into and out of the enzyme-bearing matrix.   Losses in activity may also be due to the involvement of essential active site side-chains in the covalent attachment, or to irreversible denaturation under the conditions of the coupling reaction.

Physical adsorption
Enzymes may often be tenaciously held to certain water-insoluble supports by non-covalent interactions.   The exact nature of the interactions is usually obscure, but adsorption offers the advantage that the enzyme is not chemically modified.   Ionic and hydrophobic interactions are probably important, since adsorption is found to be dependent upon pH and temperature, solvent polarity and ionic strength.

Typical water-insoluble supports for physical adsorption include alumina, glass, clays, cellulose and ion-exchange resins. The activity of adsorbed enzymes sometimes approaches that of the native enzyme in solution, but is usually considerably smaller, and as in other attachment methods success is a hit-or-miss affair. Adsorption has the advantage of simplicity, and can often be reversed by changes in the pH or ionic strength of the surrounding medium. Against this must be set the disadvantage of enzyme 'leakage' during catalysis. Slow desorption of the immobilised enzyme will both contaminate the product and progressively deactivate the catalyst.

Entrapment within polymeric lattices
In this method, enzymes are physically immobilised inside the three-dimensional lattice of highly cross-linked, water-insoluble polymers. The enzyme molecules are too large to diffuse out into the surrounding water, but small substrates and products can come and go with relative freedom. Trapping is achieved either by crosslinking of a pre-formed linear polymer in the presence of the enzyme, or by including enzyme in the original polymerisation mixture.

The advantages of this approach lie in its simplicity and the ease with which the mechanical and physical properties of the gel can be varied by suitable choice of monomer concentrations. On the other hand the activity of enzymes immobilised by this method is usually low, and there are considerable problems of enzyme leakage. A related technique, where enzymes are trapped inside synthetic fibres, is claimed to give better yields of trapped enzyme, with enhanced stability in some cases. This is a continuous process, in which water-insoluble filaments are produced by extrusion of a suitable polymer, dissolved in an organic solvent, into a water bath which contains the enzyme. Enzyme molecules are trapped in water droplets inside the filaments, which can then be woven into cloth-like fabrics or used as column materials.

Semi-permeable membranes
A new technique of great potential importance is micro-encapsulation, which involves the immobilisation of enzymes within small, spherical capsules which have a semi-permeable membrane. The enzyme molecules are larger than the mean pore diameter of the membrane and cannot diffuse out into the external solution, while smaller substrates and products can readily diffuse in and out. Most experiments with this technique have used microcapsules in which the semi-permeable membrane is an inert polymer such as nylon or polystyrene.

An advantage of this immobilisation technique is that it provides the large contact surface area between enzyme and substrate which is a feature of more established methods, but in a very much smaller total volume. Such semi-permeable microcapsules are currently being used to construct miniature artificial kidneys containing suitable enzymes and adsorbents. Enzymes successfully immobilised in this way show only minor differences in their catalytic properties, apart from the restriction on the size of substrates. The major limitation arises from the instability of many enzymes of interest when exposed to organic solvents and detergents during the preparation of the capsules.

In summary, a wide range of methods is now becoming available for the immobilisation of enzymes. The mechanisms of immobilisation are still poorly understood since the effects of adsorption, covalent bond formation, and lattice entrapment are not easily distinguished, and frequently more than one effect will contribute. More seriously, it is not yet possible to predict which immobilisation method will work best for a given enzyme, and many attempts must generally be made before a satisfactory procedure is developed.

Enzyme isolation

Pure enzymes are indispensable to fundamental studies of enzyme action but, for many practical purposes, it is better to avoid carrying out expensive and time-consuming purification procedures. Certain extracellular enzymes, including proteases and carbohydrate-degrading enzymes, are traditionally prominent in industrial applications because they are relatively stable and require only moderate purification before use. For intra-cellular enzymes, a new approach is illustrated by a recently established industrial process for production of the amino acid L-aspartic acid. The enzyme involved, aspartase, is not purified at all from its bacterial source. Instead, entire bacterial cells are immobilised by entrapment in polyacrylamide gel. When fumaric acid and ammonia are passed through columns of the gel, the aspartase in the cells catalyses the formation of L-aspartic acid (*Fig. 47*): under these conditions, aspartase retains half of its original activity even after two months of continuous operation.

Since many other enzymes of potential importance do require

FIG. 47. The aspartase-catalysed reaction.

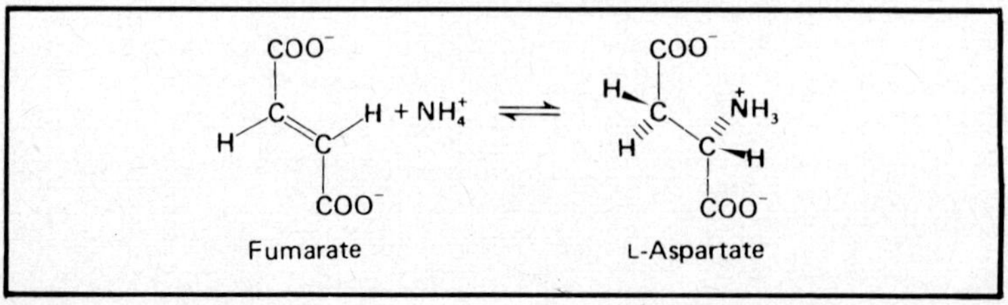

Fumarate   L-Aspartate

considerable purification before use, there is considerable interest in new, more efficient isolation methods. A conventional isolation procedure usually involves several stages, arranged in increasing order of sophistication. The first step may simply involve heating the crude tissue or bacterial extract until the stability limit of the target enzyme is all but reached, and then rapidly re-cooling the mixture. Many other proteins will be thermally denatured by this treatment, and precipitate from solution, to be removed by filtration or centrifugation. Proteins also differ in their stability to extremes of pH, and a brief exposure to low pH may lead to precipitation of unwanted protein while the target enzyme remains (relatively) intact. A milder separation method, used extensively at an early stage in enzyme purification, is based on the differential solubility of proteins in the presence of high concentrations of salts such as ammonium sulphate, or of water-miscible organic solvents.

The final steps in the purification usually involve fractionation of proteins by chromatography. Weakly basic and weakly acidic ion-exchange resins have been specially developed for column chromatography of enzyme proteins. Careful control of the pH and ionic strength of eluting buffers can give good resolution of proteins which differ either in net charge, or in their surface charge distribution. Proteins may also be separated on the basis of differences in size; the protein mixture is applied to a column containing particles of a polysaccharide or polyacrylamide gel and flushed through with a suitable buffer. The pore size of the gel beads is chosen so that the smaller proteins can penetrate into the interstitial volume, while larger molecules are excluded. The result is that large proteins emerge from the column first, and other proteins usually follow in order of decreasing size.

Unfortunately, such chromatographic procedures are rather time-consuming, and a great deal of exploratory work is often required to establish optimum conditions for resolution and enzyme recovery. Newer methods such as gel electrophoresis, where separation depends largely on differences in net charge, require costly equipment, and only small amounts of enzyme can be handled. The usefulness of most of these techniques is confined to small-scale laboratory procedures where the homogeneity of the final preparation is a more important consideration than yield.

Many hopes for a large-scale supply of purified enzymes rest with the technique called *affinity chromatography*. The basic idea of this purification method is to exploit the ability of an enzyme to recognise and bind to specific ligands, either substrates, competitive inhibitors, or co-enzymes. One of these ligands is covalently linked to an inert water-insoluble support, and the crude protein mixture containing the desired enzyme is poured onto the column. When the column is subsequently washed, extraneous protein is immediately eluted, but the target enzyme is retained on the column by virtue of its specific binding to the covalently-attached ligand. The enzyme can then be recovered either by elution with a second specific ligand, or by suitable changes in the pH and ionic strength of the eluting buffer.

The advantages of affinity chromatography are most visible when applied to the isolation of enzymes present in small amounts in the cell, and which require, therefore, extensive purification. For example, riboflavin synthase, which catalyses the final step in the biosynthesis of Vitamin B_2, has been purified 1000-fold by affinity chromatography in a single step, with a yield of 87 per cent. This should be compared with the 10-fold purification that one can expect from a conventional ion-exchange column run under ideal conditions. On a larger scale, other notable achievements include the isolation of a protease used in cancer chemotherapy, carboxypeptidase G-1, after a 35-fold purification on a single affinity column. It should be said that the parameters governing enzyme–ligand binding are extremely complex, and many other attempts at affinity chromatography have not been so successful. Nevertheless, this technique offers a rational approach to enzyme isolation, and its potential speed and resolving power continue to make it attractive for large-scale enzyme isolation.

Enzyme stabilisation

A besetting problem of enzyme technology is the inherent instability of enzymes as catalysts. Once isolated, the enzyme protein tends to denature and lose activity on storage. This activity loss is usually irreversible, and is influenced by a number of environmental factors— temperature, pH, ionic strength, solvent polarity, the presence of traces of heavy metal ions. The relative importance of these factors will change from one enzyme to another, so that such generalisations as can be made have only limited predictive value.

One interesting solution to the problem of heat denaturation has been suggested by the discovery of microorganisms which grow at 50 °C and above. It appears that many of the enzyme proteins of these so-called *thermophiles* are intrinsically more heat-stable than the proteins in species which live at more moderate temperatures. The effect of increasing temperature on the activity of a typical enzyme is shown schematically in *Fig. 48*. At 30 °C, the enzyme is sufficiently stable so that product is formed at a constant rate, proportional to the concentration of enzyme, over the period of measurement. At higher temperatures, the initial rate of product formation increases, reflecting a more rapid turnover, but the rate subsequently falls off as the protein is progressively denatured. Frequently, the denatured protein chains aggregate to give an

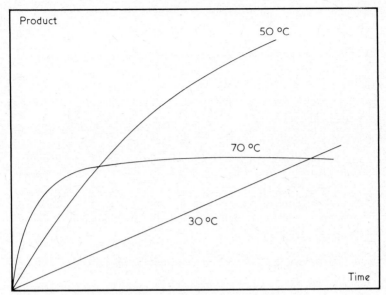

FIG. 48.   The dependence of enzyme activity on temperature.

insoluble mass, rather as egg-white proteins coagulate when an egg is boiled.

The special heat stability of enzymes isolated from thermophilic organisms may be exploited in two ways: the reaction may be conducted at a higher temperature than otherwise possible, in order to perturb an otherwise unfavourable equilibrium towards product formation, or (more usually) the heat-stable enzyme may be substituted for the enzyme from more conventional sources, in order to prolong the effective life of the catalyst.

Enzymes which contain easily-oxidised cysteine residues are generally stabilised by the exclusion of atmospheric oxygen and the presence of a suitable reducing agent.   Other additives, such as metal-chelating reagents and protease inhibitors, often have beneficial effects on enzyme stability, particularly during enzyme purification processes.

Chemical modification has also been tried as a means of enhancing the stability of enzymes.   The most common approach has been to reduce the flexibility of the protein by introducing new covalent cross-links; either between amino acid residues in the same molecule or between residues on different molecules.   After treatment with such a reagent, ribonuclease A retains half its original activity even after 30 h at 65 °C, whereas the untreated enzyme loses half its activity within 2 h at this temperature.   The disadvantage of such methods is clearly the difficulty of ensuring that the covalent cross-links do not 'lock' the protein into an inactive conformation.

Examples of specific applications of enzymes

Analytical methods

Ultraviolet and visible spectrophotometry is a convenient technique for the quantitative determination of organic species in aqueous solution, but many compounds of interest do not themselves possess a distinctive absorption spectrum. Enzymes usefully extend the scope of the analytical method by converting such compounds into more easily determined products. A typical example is the assay of ethanol by enzymatic oxidation with alcohol dehydrogenase and excess NAD^+. The reduced form of the cofactor, NADH, is produced and this is measured by its strong uv absorbance at 340 nm. This method allows the determination of ethanol concentration as low as a few parts per million.

$$CH_3CH_2OH + NAD^+ \overset{LADH}{\rightleftharpoons} CH_3CHO + NADH + H^+$$

Enzymes have also been used to extend the scope of many other analytical methods, but their relatively high cost has often discouraged their use in routine applications. Fortunately, the development of immobilised and hence re-usable enzyme preparations has partly met this objection. A particularly promising technique involves the combination of an immobilised enzyme and an ion-selective electrode. In such an electrode, the potential produced is a logarithmic function of the concentration of that ion in the surrounding solution. The most familiar of such electrodes is the glass pH electrode, but other electrodes, now commercially available, are fairly selective for, say, monovalent cations.

L-amino acids have been determined, for example, by a cation-sensitive electrode, wrapped in a cellophane membrane which immobilises a solution of L-amino acid oxidase around the electrode. The substrate diffuses through the semi-permeable membrane and is oxidised by the enzyme, producing, *inter alia*, ammonium ions (*see Fig. 21*). The change in ammonium ion concentration, in turn, is directly sensed by the cation-selective electrode. The importance of this technique is that it combines the sensitivity and specificity of immobilised enzyme reactions with the speed and experimental convenience of potentiometric measurements.

Enzymes in clinical diagnosis

The measurement of levels of enzyme activity in body fluids has become increasingly important in clinical practice; well over 10 million such tests were carried out in 1976. Such information can indicate whether disease is present even at its earliest stage, and it can often be a helpful pointer to the nature and location of the disease. Only about 15–20 enzymes have so far been widely used in this way, the majority of these being intracellular enzymes whose activity levels in blood serum are normally low in healthy individuals. A well-studied example of the use of such measurements is in the differential diagnosis of heart disease (myocardial infarction) and pulmonary embolism. In the former condition the serum levels of the transferase enzymes glutamate oxaloacetate transaminase (GOT) and creatine phosphokinase (CPK) are raised in almost all cases. In pulmonary embolism, although other symptoms may parallel those for infarction, the levels of GOT and CPK in serum remain normal.

Another useful approach is made possible by the widespread existence of *iso-enzymes*. As already noted in Chapter 1, these are enzyme proteins which differ in one, or a few, of their constituent amino acid residues. Lactate dehydrogenase, for example, which catalyses the reduction of pyruvate, in the presence of the redox cofactor NADH (*Fig. 49*) is a tetrameric enzyme, in which the constituent polypeptide chains may be of two kinds (designated M and H). Apart from M_4 and H_4 tetramers, three hybrid iso-enzymes can be constructed from the two types of subunit.

Electrophoresis of human lactate dehydrogenase shows that different tissues contain different amounts of each iso-enzyme. Heart and kidney tissues contain largely H_4 tetramers, while the M_4 iso-enzymes predominate in liver and skeletal muscle. Since these iso-enzymes differ substantially in heat stability and substrate specificity, the likely source of any unusual amount of lactate dehydrogenase in serum can be traced.

Enzymes as industrial catalysts

Of the six major classes of enzyme recognised by the International Commission, the hydrolases have received the greatest attention as catalysts for large-scale chemical applications. Most are protein-degrading enzymes, such as rennin, which is used to coagulate milk during cheese production; the bacterial proteases used as detergent

FIG. 49. The reduction of pyruvate catalysed by lactate dehydrogenase.

$$NADH + \underset{\underset{COO^-}{|}}{\overset{\overset{CH_3}{|}}{C}}{=}O + H^+ \rightleftharpoons H{-}\underset{\underset{COO^-}{\|}}{\overset{\overset{CH_3}{\|}}{C}}{-}OH + NAD^+$$

Pyruvate   L-Lactate

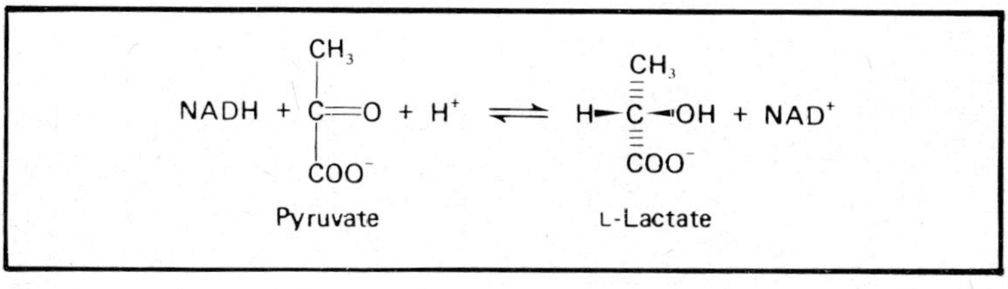

additives; and trypsin, used to soften leather. These are extracellular enzymes, cheap and relatively accessible, which require no cofactors. A recent success in enzyme technology involves the intracellular enzyme glucose isomerase, which interconverts D-glucose and D-fructose and, again, requires no added cofactors (*Fig. 50*). Glucose itself is not nearly as sweet as the equilibrium mixture, so the enzyme-produced 'high-fructose syrup' is increasingly important in the food industry.

In one process, the enzyme is immobilised within the microbial cells which produce it, by careful heat treatment. Alternatively, the enzyme may be extracted and then immobilised by adsorption onto an ion exchange resin. Exposure of D-glucose to the enzyme, under suitable conditions of temperature and pH, yields a mixture containing approximately 40 per cent D-fructose. The immobilised enzyme is recovered by filtration or centrifugation and re-used. Several million tons of high-fructose syrup are now produced annually using this method. The D-glucose starting material is obtained by the hydrolysis of corn starch, by α-amylase and glucoamylase acting in tandem, and attempts are now being made to immobilise these enzymes.

The special usefulness of enzymes as chiral reagents is shown by the optical resolution of α-amino acids through the action of amino-acylase on their *N*-acetyl derivatives. In each case, only the L-enantiomer is hydrolysed (*Fig. 51*). The pure L-amino acid is easily isolated from the produce mixture and the D-enantiomer can be obtained by subsequent chemical hydrolysis of the derivative. For larger-scale resolution of amino acids the enzyme is immobilised by adsorption onto an ion-exchange resin. Since the amino-acylase activity only starts to decay after one month at 50° C, the entire process can be fully automated.

In a similar application, tyrosinase is used to convert L-tyrosine stereospecifically into 3,4-dihydroxyphenylalanine (L-DOPA) which

Fig. 50.   The reaction catalysed by glucose isomerase.

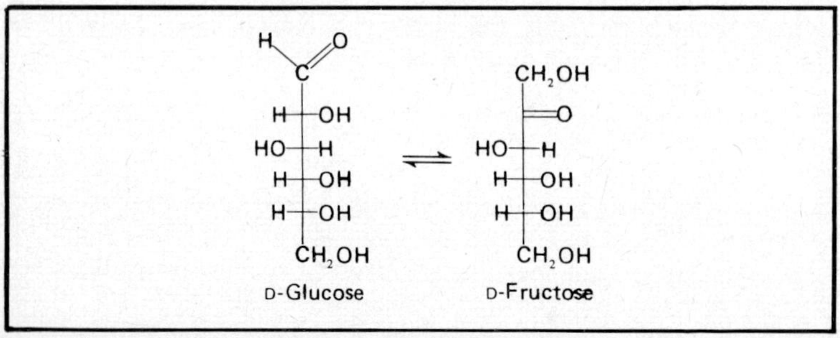

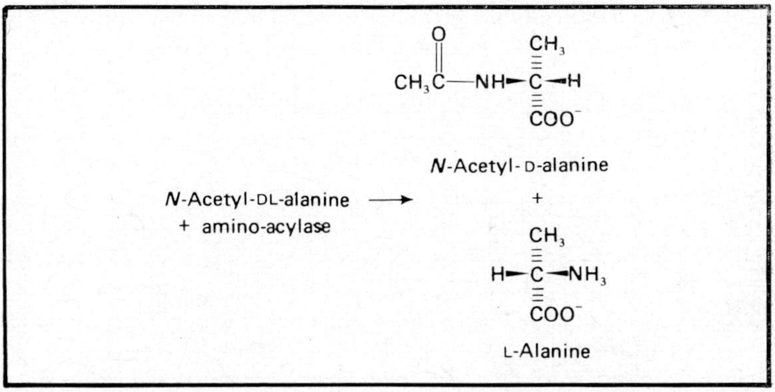

$$CH_3\overset{\displaystyle O}{\overset{\displaystyle \|}{C}}-NH-\overset{\displaystyle CH_3}{\overset{\displaystyle \|}{C}}-H$$
$$\overset{\displaystyle |}{COO^-}$$

N-Acetyl-D-alanine

N-Acetyl-DL-alanine ⟶ +
+ amino-acylase

$$H-\overset{\displaystyle CH_3}{\overset{\displaystyle \|}{C}}-NH_3$$
$$\overset{\displaystyle |}{COO^-}$$

L-Alanine

FIG. 51. Optical resolution of amino acids using amino-acylase. As shown here for alanine, the enzyme specifically hydrolyses the L-enantiomer.

is important for the treatment of Parkinson's disease (*Fig. 52*). The use of racemic (*ie* DL-) DOPA from chemical synthesis leads to undesirable side-effects.

Future developments

Much of this Chapter has been concerned with practical difficulties in using enzymes. Nevertheless, as the examples show, enzyme technology is already making an important and expanding contribution in such widely different applications as diagnostic medicine and the industrial production of sugar. Many other processes and applications are being actively investigated, with particular emphasis on an outstanding problem, the discovery of cheap and efficient ways of regenerating the expensive co-enzymes that many enzymes need for their activity.

Clearly, the biggest short-term impact will be in the synthesis of fine chemicals and drugs, where the specificity and mildness of enzyme catalysts justify their use, and in analytical applications

FIG. 52. The specific hydroxylation of L-tyrosine by tyrosinase.

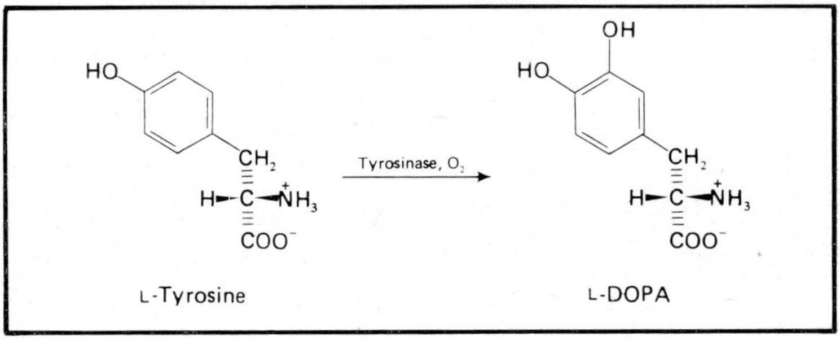

L-Tyrosine   L-DOPA

where only small amounts of catalyst are required. Enzyme research has other benefits too — much of the interest of the brewing industry is aimed less at the development of new processes and products than at maintaining finer control over traditional ones.

The exploitation of enzymes as industrial catalysts has only just begun. Any new process using enzymes has to be thoroughly tested before it can be relied upon to compete effectively with established conventional methods. There is, though, a great potential for novel enzyme-based processes. The sort of problem which enzymes might tackle in the future is illustrated by the hydrolytic enzyme cellulase, under active development in the US. This enzyme, isolated from certain microorganisms, catalyses the breakdown of wood cellulose into glucose, which can be fermented further into ethanol or acetone. An industrial process for converting waste cellulose into food or industrial solvents (presently obtained from petroleum-based products) may not be far away.

Appendix

Hints on viewing stereo diagrams

The three stereo diagrams in this monograph can be viewed with the naked eye as follows: with the page flat and well-illuminated, and eyes at normal reading distance, allow the eyes to relax. Three blurred images will be seen. Concentrate on the central one until it comes into focus as a three-dimensional image.

This technique is not easy to master, and a magnifying stereo viewer is recommended, preferably one with the distance between the eyepieces adjustable. Even using this aid, it normally takes a little time to become accustomed to viewing stereo pair diagrams. One technique is to focus attention on a particular feature of the diagram, and to try to bring the two versions into coincidence. If this succeeds, the observer then becomes aware that certain parts of the diagram 'stand out' from the page. It is possible, for example, to tell whether a helix is wound in a right-handed (clockwise) or a left-handed (counter-clockwise) sense.

Viewers can be obtained from, *eg*, Capital Biotechnic Developments, 66A Churchfield Road, London W3, or Duval Studios, 217 High Road, Chiswick, London W4 2DP.

Suggestions for Further Reading

Articles
There are relatively few articles on enzymes that are written for the non-specialist, but the following are all useful:

Scientific American
D. C. Phillips, 'The three-dimensional structure of an enzyme molecule', Nov. 1966, **215**, 78.
K. Mosbach, 'Enzymes bound to artificial matrices', March 1971, **224**, 26.
F. Wroblewski, 'Enzymes in medical diagnosis', Aug. 1961, **205**, 99.
R. M. Stroud, 'A family of protein-cutting enzymes', July 1974, **231**, 74.

Chemical Society Reviews
W. N. Lipscomb, 'Three-dimensional structure and chemical mechanism of enzymes', 1972, **1**, 319.
C. J. Suckling, 'Immobilised enzymes', 1977, **6**, 215.

Books
Almost all biochemistry texts contain sections on enzyme chemistry. Three of the most up-to-date specialised books on enzymes are:

W. A. Ferdinand, *The enzyme molecule*. New York: Wiley, 1976.
A. Fersht, *Enzyme structure and mechanism*. Reading: Freeman, 1977.
K. G. Scrimgeour, *Chemistry and control of enzyme reactions*. London: Academic, 1977.

Other books covering selected topics are:

Experiments for school use
D. G. Mackean, *Experimental work in biology No 2 (enzymes)* (with teacher's supplement). London: Murray, 1971.

Protein structure
R. E. Dickerson and I. Geis, *The structure and action of proteins*. New York: Harper and Row, 1969.

Chemical mechanisms
W. P. Jencks, *Catalysis in chemistry and enzymology*. New York: McGraw-Hill, 1969.
J. N. Lowe and L. Ingraham, *An introduction to biochemical reaction mechanisms*. Englewood Cliffs, NJ: Prentice-Hall, 1974.

Glossary

Active site, the portion of an enzyme surface directly involved in substrate binding and catalysis.
affinity labelling, use of an inhibitor structurally similar to an enzyme substrate to effect specific chemical modification of an active site.
assay, analysis for enzyme activity.

Chirality, a molecule possesses chirality if it exists in either of two forms related as an object and its non-superimposable mirror image.
coenzyme, a cofactor that is readily dissociable from an enzyme.
cofactor, a metal ion or small organic molecule required for enzyme activity.
configuration, the three-dimensional structure of a molecule, determined by the constituent atoms and the way they are bonded to one another. If a molecule can exist in two or more non-interchangeable configurations, these forms are called stereoisomers.
conformation, for a given configuration, one of a (usually large) number of spatial arrangements readily interconvertible by rotation around single bonds.

Denaturation, a gross change from the active (native) conformation of an enzyme protein, paralleled by a loss of catalytic activity.

Enantiomers, stereoisomers mutually related as object and non-superimposable mirror image.
enzyme, protein molecule that catalyses a specific chemical reaction.

Immobilisation, confinement of an enzyme by physical or chemical means.
isoenzymes, enzymes from the same organism that differ only marginally in their primary structure and catalytic activity.

Polypeptide, straight-chain polymer formed by condensation of amino acid units.
primary structure, the characteristic sequence of amino acids in a particular peptide or polypetide chain.
prosthetic group, a coenzyme bound tightly (and often covalently) to an enzyme active site.

Quaternary structure, the existence of more than one separately-folded polypeptide chain within a larger enzyme unit.

Reaction coordinate, often expressed as a function of certain bond lengths or angles, it measures the progress of a reacting system towards product formation.

Secondary structure, refers to the existence of regularities in parts of the backbone conformation of most, if not all, enzyme proteins.
stereoisomers, forms of a molecule differing only in the spatial distribution of atoms and functional groups.

C-terminus, the end of a peptide chain containing a free α-carboxyl group.

N-terminus, the end of a peptide chain containing a free α-amino group.

tertiary structure, the organisation of an enzyme polypeptide chain in three dimensions, including details of the disposition of amino acid side-chains.

thermophile, organism adapted to live at an unusually high ambient temperature.

transition state, the highest point in the free energy profile for a reaction. The energy difference between the starting materials and the transition state represents the minimum activation energy required for reaction to occur.

turnover number, expresses the catalytic activity of a pure enzyme as the number of substrate molecules converted to product by each enzyme molecule in unit time under defined assay conditions.